I0795122

Aircraft of the Chaco War

Aircraft of the Chaco War

Dan Hagedorn and Antonio Luis Sapienza

Schiffer Military/Aviation History
Atglen, PA

ACKNOWLEDGEMENTS

The authors are indebted to many persons for their generous contributions and who have assisted with photographs and the fruits of their own research, and without whose unselfish assistance this book would have been far less complete.

Contributors (Bolivia)
Sgto. Ramiro Molina Alanes, FAB; Coronel Av.(R) Amalia Villa de la Tapia; Coronel Victor Maldonado G.

Contributors (Paraguay)
Coronel Av.(R) Agustin Pasmor; Tte.Obs.(R) Gonzalo Palau; Capt.P.A.M.(R) Abdon Alvarez Albert; Coronel P.A.M.(R) Enrique Dentice, FAP; the Fuerza Aerea Paraguaya; the Aviación Naval Paraguaya and the Ministerio de Defensa Nacional; special thanks to the staff of the Museo de Historia Militar

Contributors (Europe)
Harm J. Hazewinkel, Eric B. Morgan, Angela M. Wooton (Imperial War Museum, London)

Contributors (U.S.)
John M. Davis, Captain Robert F. Dreeson, USN (Ret), Staff Volunteer at the National Air and Space Museum Branch Library, Dr Gary Kuhn, to whom a very special debt of gratitude is extended in connection with his editing of the original manuscript.

Book Design by Ian Robertson.

Copyright © 1997 by Dan Hagedorn & Antonio L. Sapienza.
Library of Congress Catalog Number: 96-70264

All rights reserved. No part of this work may be reproduced or used in any forms or by any means – graphic, electronic or mechanical, including photocopying or information storage and retrieval systems – without written permission from the copyright holder.

Printed in China.
ISBN: 0-7643-0146-2

We are interested in hearing from authors with book ideas on related topics.

Published by Schiffer Publishing Ltd.
77 Lower Valley Road
Atglen, PA 19310
Phone: (610) 593-1777
FAX: (610) 593-2002
Please write for a free catalog.
This book may be purchased from the publisher.
Please include $2.95 postage.
Try your bookstore first.

Printed in Hong Kong.

CONTENTS

Acknowledgements **4**
Introduction **6**
Historical Background **10**

Chapter 1 Opening Hostilities ***16***

Chapter 2 Bombers, Reconnaissance, and Multi-Purpose Aircraft ***18***
Curtiss-Wright A14R "Bolivian Osprey" (Bolivia) 18
Potez 25 A-2 and Potez 25T.O.E. (Paraguay) 29
Vickers Type 149 "Vespa III" (Bolivia) 42
Junkers K 43b (Bolivia) 45
Curtiss "Cyclone Falcon" (Bolivia) 49
Breguet 19 (Bolivia) 54
Fokker C-Vb (Bolivia) 58
Curtiss D-12 "Falcon" (Paraguay) 60
Macchi M.18A.R. (Paraguay) 61
de Havilland/ADC D.H.9 (Bolivia) 65
Ansaldo SVA-5 and SVA-10 (Paraguay) 65

Chapter 3 Fighters and Fighter Bombers ***66***
Curtiss Model 35A "Hawk II" and Model 65
"Sea Hawk II" (Bolivia) 66
Fiat CR.20bis (Paraguay) 73
Vickers Type 143 "Bolivian Scout" (Bolivia) 79
Wibault Type 73 C.l (Paraguay) 82
Savoia S.52 (Paraguay) 86
Gourdou-Leseurre LGL 32 C.l (Bolivia) 89

Chapter 4 Other Types—Transports, Ambulances, and Trainers **90**

Transports **90**
Junkers F 13L and F 13W (Bolivia) 90
Junkers W 34ci (Bolivia) 94
Junkers Ju 52/3mde and Ju 52/3mge (Bolivia) 96
Travel Air Model S-6000-B (Paraguay) 101
Breda 44 (Paraguay) 103
Savoia S.59bis (Paraguay) 105
Ford Model 5-AT-D (Bolivia) 106
Curtiss "Robin C" (Paraguay) 107
Breda 15S (Paraguay) 107
Consolidated 21-C (Paraguay) 108
Waco Model C (Paraguay) 110
Cant 10ter (Paraguay) 110
Sikorsky S-38B (Bolivia) 111

Bolivian Trainers **112**
Caudron C97 112
Caudron C59Et2 114
Vickers Type 155 "Vendace III" 115
Morane-Saulnier 35 E.p.-2 and 139 116
Curtiss-Wright CW-16E "Trainer" 118
Junkers A 50 "Junior" 119

Paraguayan Trainers **120**
Fleet Model 2 120
Junkers A.50fe "Junior" 122
Hanriot H.D 32E p.2 123
de Havilland D.H.60G and D.H 60M "Moth" 124
Morane-Saulnier 35E.p.-2 and 139E.p.-2 125
Cant 26 128
S A M L. A.3 (S-l) 128

Chapter 5 Things That Didn't Happen **129**
Chapter 6 The Aftermath **137**
Sources **143**

Introduction

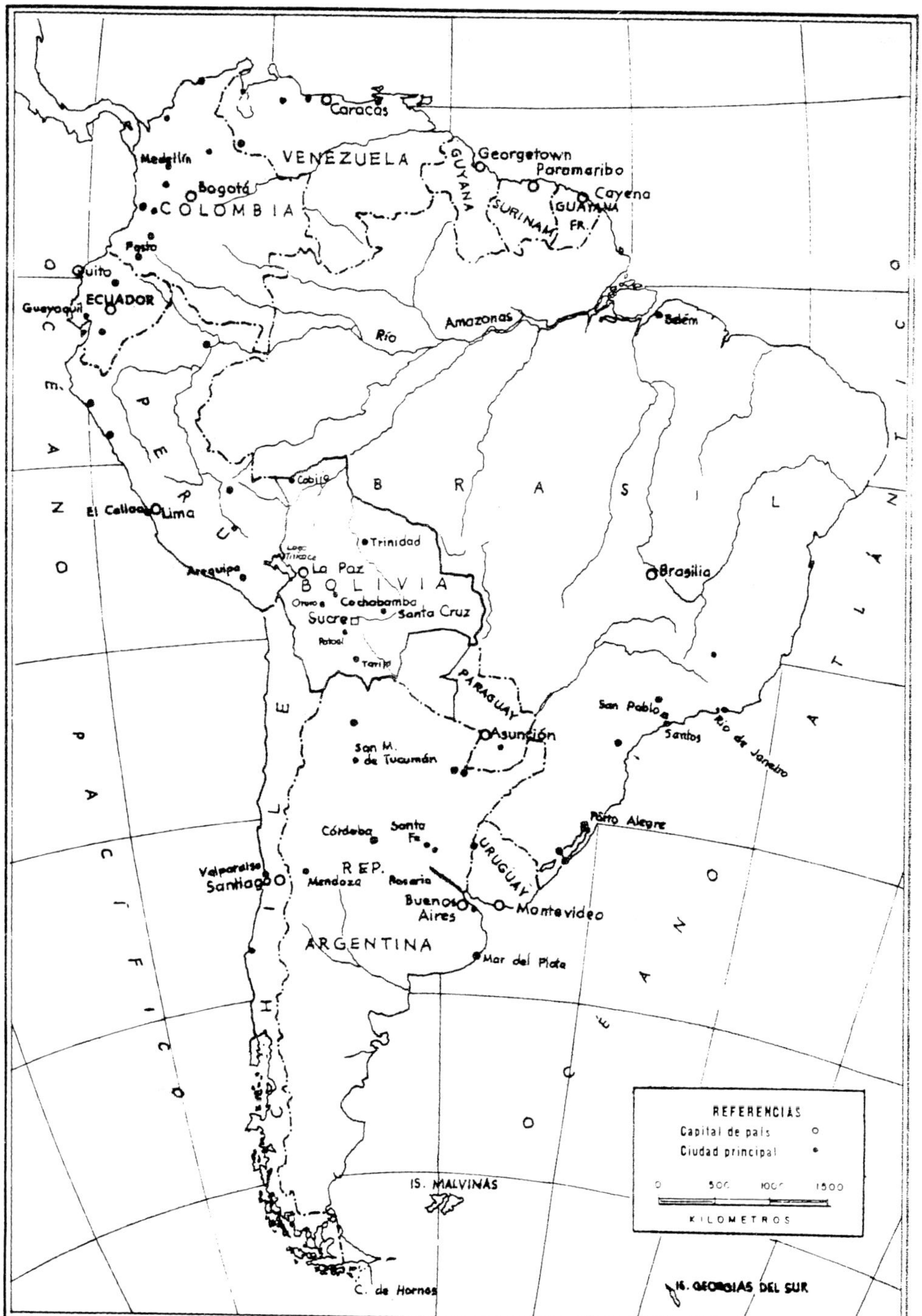

The South American land mass, showing the relative size and locations of Bolivia, Paraguay and, sandwiched in between, the Chaco Boreal.

Usually known outside of Bolivia and Paraguay as the Chaco War, this was a very costly conflict between these two nations, during the period of most intense hostilities, between 1932 and 1935, but having incidents of open conflict as far back as 1928.

Fought over the Chaco Boreal, or Gran Chaco, a vast wilderness region of about 100,000 square miles (260,000 square kilometers) north of the Rio Pilcomayo and west of the Rio Paraguay, this conflict is of interest to aero historians in that it was the first modern conflict in the western hemisphere, fought between two established nations, in which aircraft and aviation played a significant part.

The conflict and its roots are complex. However, it may be fairly stated that it stemmed from the much earlier outcome of the War of the Pacific (1879-1884) in which Chile defeated Bolivia and annexed its entire Pacific coastal region. Thereafter, Bolivia made it almost a matter of national pride to break out of its landlocked circumstances through the Rio de La Plata system to the Atlantic coast. Unfortunately, athwart that route lay the Gran Chaco region, which, aided by the not so subtle suggestions of foreign entrepreneurs, the Bolivians had been led to believe might contain vast oil reserves as well. The war

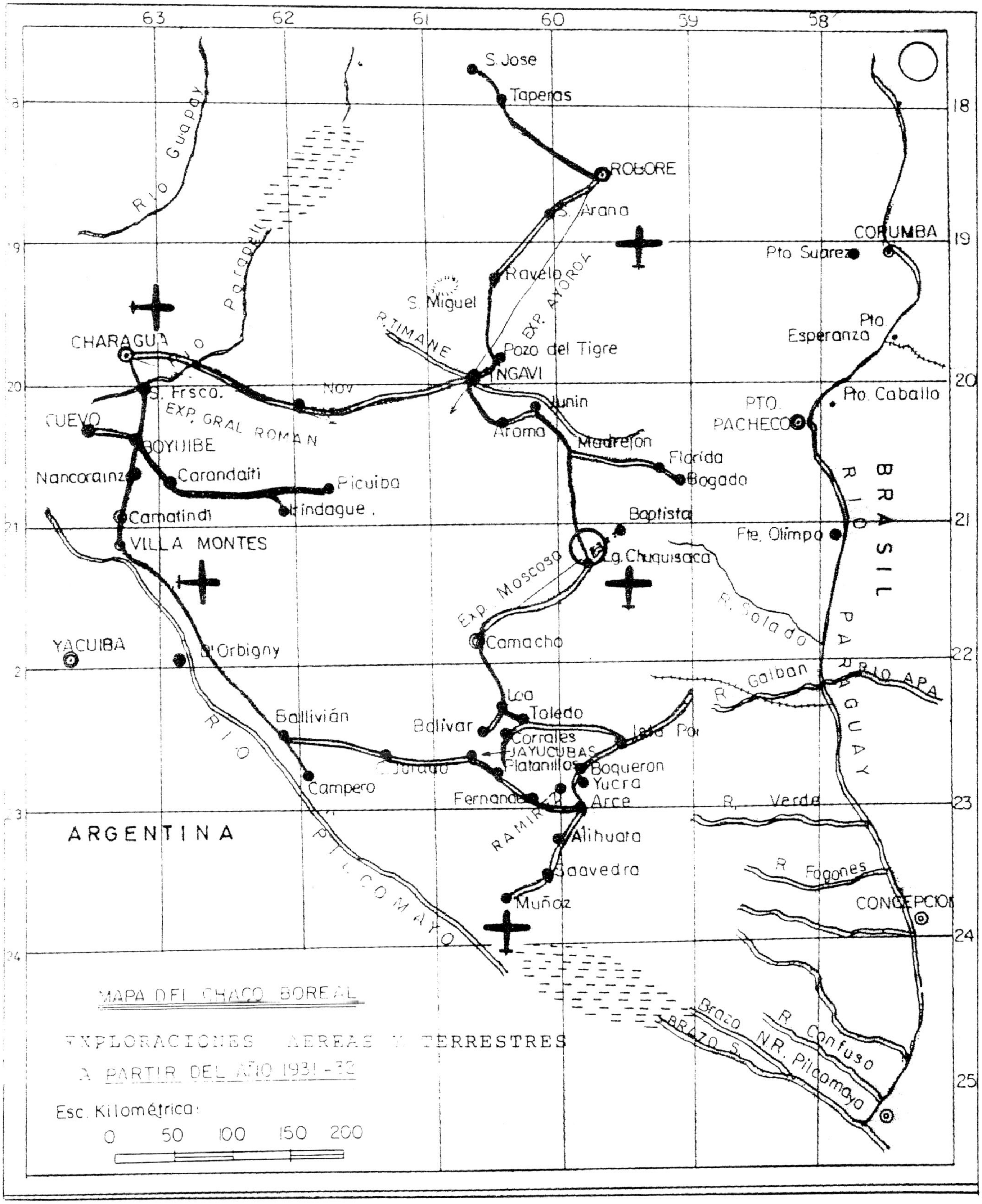
63
62
61
60
59
58
18
19
20
21
22
23
24
25
Rio Guapay
Parapeti
S. Jose
Taperas
ROBORE
S. Arana
CORUMBA
Pto Suarez
Ravelo
S. Miguel
EXP. AYOROA
R. TIMANE
CHARAGUA
Pto. Esperanza
Pozo del Tigre
INGAVI
Nov
S. Frsco.
EXP. GRAL ROMAN
Junin
Pto. Caballo
PTO. PACHECO
Aroma
Madrejon
Florida
Bogado
BOYUIBE
Carandaiti
Nancorainza
Picuiba
Lindague .
Camatindi
VILLA MONTES
Baptista
Fte. Olimpo
BRASIL
RIO
Lg. Chuquisaca
Exp. Moscoso
R. Salado
PARAGUAY
YACUIBA
D'Orbigny
Camacho
R. Galban
RIO APA
Toledo
Ballivián
Bolivar
Corrales
JAYUCUBAS
Platanillos
Boqueron
Yucra
Arce
Campero
Fernandez
R. Verde
RIO
ARGENTINA
RAMIREZ
Alihuata
PILCOMAYO
Saavedra
R. Fogones
Muñoz
CONCEPCION
Brazo
BRAZO S.
N.R. Pilcomayo
R. Confuso
MAPA DEL CHACO BOREAL
EXPLORACIONES AEREAS Y TERRESTRES
A PARTIR DEL AÑO 1931 - 32
Esc. Kilométrica:
0 50 100 150 200

has often been described, by foreign observers and historians, as a "limited" war for Bolivia, who saw the Chaco as merely a desirable acquisition with exploitation possibilities while, for Paraguay, it was literally a matter of survival.

On the surface of it all, Bolivia would have seemed, at the time, to hold a virtually overwhelming advantage over Paraguay: it had easily three times the latter's population, a standing Army well trained by the German General Hans von Kundt, and the wherewithal to secure an ample supply of arms - including aircraft - bought largely via loans secured from United States banks in league with the oil barons and speculators.

But the general morale of Bolivia's Army, consisting in the main of Indian draftees, was low, and the underdog Paraguayans were far better suited to fight in the lowland swamps and jungles that constituted the Chaco region, in which many of the hapless Bolivians died of disease and snakebite, as well as gunfire.

Although both nations had maintained small military garrisons in the region, it is general recognized that, on 5 December 1928, Paraguay initiated a series of clashes (due to Bolivian invasion of Paraguayan territory according to Paraguayan sources), which eventually led to full-scale war on a surprising scale. Both sides moved ever more troops into the region and, by 1932, all-out war was definitely underway.

In June 1932, the Bolivians seized Paraguayan strong points in the northern Chaco and launched a successful attack in the central Chaco against Fortín Boquerón. This prompted Paraguay to order a General Mobilization in August, and to dispatch a force under Lieutenant Colonel José Estigarribia in their first major offensive to retake Fortín Boquerón, which finally fell, after a heroic defense, at the end of September. General Kundt was recalled to La Paz as a result of this reverse, but retained overall command, organizing a concentration of Bolivian forces in the south to attack Fortín Nanawa, where there was very heavy combat for several months.

Paraguay formally declared war on Bolivia on 10 May 1933, and neutrals intensified largely futile efforts to mediate an end to the by now all-out war. Colonel Estigarribia launched a series of Paraguayan attacks along a very extended front late in October 1933 and made such impressive gains that the Bolivian President, Daniel Salamanca, dismissed the Bolivian leadership from command of the front and replaced him with General Enrique Peñaranda. At the end of a three-week truce (which gave the Bolivians, in retrospect, an invaluable breathing space) Estigarribia renewed the Paraguayan drive on 9 January 1934, with a major attack against the key Bolivian outpost at Ballivián on the Rio Pilcomayo where, from March into July 1934, the heaviest combat of the war occurred. Ballivián finally fell on 17 November 1934, and President Salamanca was forced to resign by an Army junta in rather unusual circumstances. Meanwhile, the Paraguayan advance continued into what was, by then, indisputably Bolivian territory, into January 1935.

OPPOSITE: General map of the Chaco region showing the principle geographic locations, airfields and primitive road system circa 1931-32 (FAB).

After frenetic and costly Bolivian counterattacks finally put Paraguayan forces on the defensive, a truce was arranged on 12 June 1935.

It is estimated that something near 100,000 men lost their lives during the war. A peace treaty was finally arranged by the Chaco Peace Conference, which included the interested parties Argentina, Brazil, Chile, Peru, Uruguay and the United States, and was finally signed in Buenos Aires on 21 July 1938. As a result, Paraguay gained clear title to most of the disputed region, but Bolivia was given a corridor to the Rio Paraguay and a port in that extremity, Puerto Casado. The war caused massive disruption of both economies, and, in the case of Bolivia, resulted in demands for reform amongst the long-suffering Indian masses. It should be mentioned that Argentina can claim the primary credit for the settlement (a Nobel Peace Prize was subsequently awarded) and, while both Paraguay and Bolivia ended up getting essentially what they had set out for, Argentine investors profited enormously from the Paraguayan territorial gain.

Although the aerial involvement in the Chaco War may, by World War standards, be viewed as rather insignificant, it should be understood at the outset that military aeronautics and theory were in their infancy in Bolivia and Paraguay, and the problems incident to acquiring, manning and fielding aircraft were, to say the least, formidable.

Although actual aerial combats were rare, this was due more to the relatively small numbers of aircraft engaged than to any lack of initiative or aggressiveness on either warring side.

Aviation did, indeed, play a significant part in the greater land conflict and, although several accounts in Spanish have been published in Bolivia and Paraguay, these have been, perforce, somewhat biased and ethnocentric, and until now, no comprehensive, overall English language account has been offered.

The scope of this account, as will be seen, focuses not only on the actual employment of aviation assets by both sides, but to their acquisition and, often, the adversities encountered in fielding what often proved to be less than satisfactory equipment.

This account will also investigate the frantic efforts made by both sides to acquire additional aircraft at various stages, including some frustrated attempts, with a word or two about the implications had these acquisitions in fact been received. Additionally, we will outline the post-armistice purchases made by both sides, in spite of the truce agreement, to gain more modern equipment should hostilities have resumed.

Throughout this account, the reader is encouraged to recall several historical conditions that contributed to the events as they unfolded. First and foremost, perhaps, was the fact that the world at large was thrust into the depths of a great economic depression at almost precisely the chronological span of the conflict. This had a predictable effect on the major aircraft manufacturing firms in both the United States and Europe who were, literally, struggling to survive. The sale of aircraft to belligerent's in this environment thus

An aerial view of the main field and birthplace of Paraguayan military aviation, Ñu-Guazu aerodrome near Asunción, as it appeared in 1932. The large building in the center still exists today. (Museo de Historia Militar via A. L. Sapienza)

became a heated contest, and amongst the earliest victims were business ethics, in the process.

The discerning student of the subject will quickly recognize that, as in most modern conflicts, the two belligerents involved in the Chaco War went to war with what they had on hand at the time; both subsequently exerted considerable efforts towards acquiring additional aircraft as events unfolded; Bolivia emerged the clear champion of such undertakings, due mainly to her vastly superior financial backing. As a result, the aircraft employed during the course of the conflict (1928-1935) underwent an almost complete transformation (in the case of Bolivia) while Paraguay fought on, for the most part, with what she started with. Predictably, both sides sought post-armistice aircraft acquisitions in secret as insurance against the other should hostilities once again break out.

As this book has intentionally been focused on the aircraft involved, it is appropriate at this point to acquaint the reader with regard to the aircraft designations used and the rationale for same. There has been considerable confusion over the years on this subject, with not a little of this fostered by the small number of publications in Spanish issued by writers in both Bolivia and Paraguay. These efforts may be forgiven for frequent errors in presentation due to the inevitable habit of the airmen themselves, who quite rightly have contributed, as in other wars, to oversimplifying references to aircraft they flew. The authors of this work have labored to clarify these contentious issues. For example, two aircraft types that featured prominently are often cited as the Breguet XIXA.2 and the Potez XXVA.2. In fact, the French themselves, as early as February 1928, were promoting these two aircraft types for export as the Breguet 19 A-2 and the Potez 25 A-2, although both had earlier been cited with the Roman numeral presentations. Whenever possible, in this account, designations will be presented as nearly correctly as possible, and the rationale for such selection will be cited.

Historical Backgroound

In terms of sheer numbers, the aviation assets available to both sides during the course of the Chaco War were really rather small.

Not counting operational and training losses, which will be referred to further on in this account, Bolivia operated somewhere between 57 and 62 combatant or combat capable aircraft between 1928 and 1935, as well as approximately 22 support and training aircraft.

Paraguay could count only about half as many combatant types, 32, but had an equal number of support types and trainers (23), which actually saw more involvement in operations than comparable types in Bolivia.

At the outset of hostilities in late 1928, military aviation in both countries was barely underway. In both instances, the entrenched military leadership took a rather dim view of aviation and, in keeping with the very strong European influences used to form their respective aviation establishments, they adopted the operational philosophies then current in France and Switzerland.

Although several half-hearted attempts had been made earlier, military aviation in Bolivia is officially credited as having commenced in August 1924 with the creation of the Cuerpo de Aviadores. The first attempt by Bolivia had been made in April 1913 when President Villazon accepted an offer made by the Rapini brothers (Miguel and Napoleón) of Chile to make flights in La Paz on the altiplano. The first test flight was made on what was described as the "newly constructed runway" on the altiplano in the vicinity of La Paz, believed to be near the location of the present day "El Alto" airport. The test, performed in a radically underpowered Bleriot XI type monoplane with a 50hp Gnome engine was unsuccessful, due to the altitude, and another attempt was not made until July 1915 when two more Chilean pilots, Luis O'Page and Clodomiro Figueroa (the latter flying a 80hp powered Bleriot XI), tried again on behalf of a government invitation at Oruro and La Paz, respectively. These attempts were only partially successful, as one of the pilots (Figueroa) was only able to attain an altitude of about 10 feet, while the attempt of Omar Page at Oruro reached about 200 meters on 22 August. According to one source, there were at least 21 unsuccessful attempts to fly at La Paz in the years to follow.

In September 1916, Bolivia sent three pilots to the new flying school at "El Palomar" in Argentina; Captain's René Pareja, José R. Alarcón and Sub Teniente Horacio Vasquez. Three other officers were sent to Chile to the Escuela Militar at "El Bosque" for instruction at about the same time; Sub Teniente's Bernardino Bilbao Rioja, Ernesto Arévalo and Alfonso Crespo. Of these, Captain Alarcón became immortalized on 23 January 1917 when he crashed a 50hp Farman trainer at "El Palomar" during a solo flight and was killed, thus becoming the first Bolivian aviator to die in line of duty. Meanwhile, in 1916, the Bolivian congress authorized the foundation of the Escuela de Aviación Militar, although in fact the school did not become established until several years thereafter. Three additional Bolivian Army officers were dispatched to Spain for training at the Escuela de Cuatro Vientos; these having been Tenientes Raúl Vargas Guzmán, Lillo Mallo and Carlos Peña.

While these young officers were being schooled abroad, the Bolivian Government decided, in light of the previous failures of various foreign aircraft, to engage two engineers residing in Argentina, Albert Jalfert (a Dane) and Andrés Tomsich (an Austrian immigrant) to build a specialized aircraft configured for the unique operating extremes at La Paz. This aircraft, named "Cóndor," was essentially a Taube design with a 100hp Argus engine. Finished around November 1917, the pilot, Jalfert, pronounced the aircraft "not airworthy" due to center-of-gravity problems (he apparently believed the engine was too heavy for the airframe) and refused to fly the aircraft until modifications had been effected by Tomsich. Tomsich, in turn, decreed that the engine was necessary for the altitude and that the aircraft was otherwise satisfactory! Eventually, the aircraft was test flown once, on 7 April 1918, by an Argentine aviator named Chumiento, who promptly crashed it. Repaired, however, the aircraft is alleged to have been flown again successfully, on 5 August by Tte. Horacio Vásquez, although only briefly.

Although bearing the LAB logo on its nose, the forward gun position is barely concealed under the tarp on Junkers Ju 86z-7 CP-ANM (serial number 3) which also bears the Bolivian tricolor on its rudders. This overall view on page 12 of CP-ANM/3 shows that the dorsal gun position on this "airliner" was clearly lurking beneath the tarp in these Cochabamba views. (Hagedorn Collection).

In 1919, Donald Hudson, a United States citizen, was commissioned by the Government and brought with him a Curtiss 18-T "Wasp" triplane. Aided by R. O. Albaugh, Hudson was largely responsible for setting the stage for the further development of aviation in Bolivia. In his own words, Hudson ..."selected a location for a field and hangar in March 1920. The field was laid out 1,000 by 1,500 meters and from this area was taken approximately 85,000 tons of rocks, averaging two pounds apiece. This work was done entirely by Indian labor at about .35 cents gold, each, per day. The cost of clearing the field and construction of the hangar amounted to about $30,000 gold. When the field had been cleared of the enormous number of loose stones and the hangar completed, the Government became rather sarcastic in their replies to requests for more money and time in which to surface the field properly." As a result, Hudson threw caution to the wind and "decided to sacrifice the safety factor that a smooth landing place would give us" and made the first test flight with the "Wasp."

Two days after the Curtiss 18-T arrived in crates (15 April 1920), it had been erected, the engine tuned, and Hudson arrived at 4:30PM "and the bus was ready for a hop." Hudson reportedly "jumped into the machine without so much as a wave,`revved' it up and within two minutes of the time of his arrival at the field, was in the air," with 65 gallons of fuel, five gallons of oil, eight gallons of water, the pilot and his observer and mechanic, Albaugh.

Taking off into a 10MPH headwind, the "Wasp" was airborne in less than 70 meters, much to the approval of the officials gathered to witness the historic flight. "We stayed up about 35 minutes, during which time we stunted over the city of La Paz" according to Albaugh, although due to the failure of the radiator shutters to stay open, the flight had to be terminated. This, the first truly successful

flight in Bolivia, from the 13,623 foot altitude of the altiplano (believed to have been the highest take-off in aviation at that time) ended with a "perfect `three-pointer' on the rough ground and the roll-out was at least 10MPH slower than a DH-4 lands at sea level" according to Albaugh. The Bolivian Minister of War was quoted as saying, upon the alighting of the duo, that "You Yanks are men of action - not words!"

The Minister of War is also quoted as stating publicly that, as a result of Hudson's spectacular flight, the Government had decided to establish and "maintain permanently, two aero squadrons of at least 18 planes," and went on to say, "this is only a starter." Unfortunately, Hudson crashed the "Wasp" on the first flight attempted between Oruro and La Paz on his return, near Sicasica, Bolivia, on 20 May 1920, by which time he had been commissioned as a Teniente Coronel in the Bolivian Army.

At the time of the loss of the "Wasp," Hudson was not the only officer in the Bolivian military aviation establishment. Teniente Coronel Victoriano Gutierrez had been named Director of the Escuela de Aviación Militar and four Bolivian pilots (possibly graduates of the Argentine training program cited earlier) were also noted as being on the faculty of the school. These were Captain's Rene Paroja and Alfonso Crespo and Lieutenants Bernardino Bilbao and Ernesto S. Arévalo.

In January 1921, the famous French pilot Mauricio Bourdon arrived in La Paz in company with three aircraft for the Bolivian Army which had been presented to the Government by Sr. Simón I. Patiño, a tin magnate, who would subsequently bequeath further aircraft to the fledgling service. The aircraft in question consisted of a Bleriot trainer with an 80hp engine, the exact variant of which remains unclear, and single examples of the sleek Spad-Herbemont S.20 with a 300HP Hispano-Suiza engine and a smaller S.34 with an 80hp LeRhone. The French pilot entertained the citizens of La Paz on 8 March 1921 in the Bleriot, which he somehow coached off the altiplano. Bourdon and a Bolivian passenger, Sr. Agustín Fernandez, secretary to Sr. Patiño, were killed sometime after this in the crash of one of the aircraft. One other aircraft arrived in Bolivia during 1921. This, an otherwise unidentified Ansaldo SVA, was brought in from Argentina by a Tte. Mendoza, but it was destroyed in a crash at his hand sometime shortly after its arrival. It is not clear if this aircraft was Bolivian Army property or not. The fate of the Spad-Herbemonts and the other early military aircraft are unknown at this writing.

The practical birth of aviation in Bolivia was greeted with enthusiasm by the public and gentry and public collections or subscriptions were made in the cities of Potosi, Oruro and La Paz for the purpose of purchasing additional aircraft. In 1922, the Government, with these funds in hand, ordered four brand-new aircraft in France, these almost certainly the four Morane-Saulnier M.S.35E.p.-2 trainers which arrived 22 July 1923. Although also described as "Morane A.R. type" with 130hp Clerget engines, these aircraft were accompanied by two "Pinguinos" (non-flyable, taxi trainers with small Anzani engines, thus the name "Penguins"). At about this same time, the re-born Escuela Militar de Aviación was again institutionalized by the Government of Dr. Bautista Saavedra on 7 September 1923 by Supreme Decree and activities once again formally inaugurated on 12 October.

The contract for these aircraft included the services of a French instructor, one Eduardo Deckert, and mechanic Andres Ceradin, for a period of one year, during which time they were to attempt to train 15 pilots and eight mechanics. For all practical purposes, the M.S.35E.p.-2s were the first "standard" Bolivian military aircraft, and they gained serials "1" to "4".

Using one of these aircraft, Captains Vargas Guzman and Pena y Pillo made the first all-Bolivian cross-country flight in August 1923, although it is not clear where these pilots gained their training to do so, as they were not previously listed on the strength of the EMA. Classes for the instruction and training of new pilots by the French mission started in earnest in January 1924, but due to "technical" and economic difficulties (at least two and possibly three of the M.S.35s had suffered damage at the hands of students, this constituting the "technical" dilemma), the school was suspended and the pilots enrolled at the time were sent to Germany, instead, for further instruction, a fact seldom reported.

With the departure of the rather lackluster French mission of Deckert, the Bolivians arranged for a Swiss Mission to further the establishment of their military aviation corps. As a result, in June 1925, Lieutenant Hans Haeberli, formerly of the Swiss Military Aviation, arrived to reopen the EMA. Although nominally under the command of Major Bernardino Bilbao at the time, Haeberli was eventually awarded similar rank and, in fact, provided the actual leadership of the school. Haeberli remained in Bolivia for five years and, during that time, was able to establish the EMA as a permanent institution. He single handedly trained not fewer than 26 pilots, and has seldom received due credit for this important contribution, one which was to materially assist Bolivia in the coming conflict with Paraguay. His first class of students graduated only two pilots, LT's Jose Coello and Faustino Rico but, on 12 October 1925, Rico made the first flight of a Bolivian graduate of its own EMA. The following year, the second class graduated nine pilots.

The aircraft acquisitions made by Bolivia under the guidance of Haeberli and its own initiatives between the time of the arrival of the M.S.35E.p.-2s in July 1923 and the actual, formal creation of the Cuerpo de Aviadores in August 1924, is somewhat puzzling. While most sources cite only the M.S.35s and the "Pinguinos" as being on strength during this period, the prestigious JANE'S All the World's Aircraft for 1926 (which of course covered the events of the preceding year, 1925) reported that the Bolivians had five Fokker C-VCs with 500hp Hispano-Suiza engines, two Fiat B.R.s with 700hp Fiat engines, two Breguet XIXA.2s, one unidentified Potez aircraft, 10 Bristol aircraft with 250hp Rolls-Royce Falcon engines (which would seem to indicate F.2B "Fighters"), not less than nine Martinsyde biplanes with 200hp Wolseley Viper engines (which would seem to indicate F.4 "Buzzard" variants) and 11 assorted Avro 504s with a combination of Monosoupape, LeRhone, Lynx and Viper engines! Of course the Fokker C-Vcs and the Breguet XIXA.2s are well-known acquisitions, but the remainder of these aircraft are undocumented, with but one exception, which tends to lend credence to the JANE'S report. A photo of what appears to be a Fiat B.R. (but more likely a similar R-2) in Bolivian markings and named "Cobija" shows up in a well-regarded history of Bolivian aviation. It appears likely that one of the Fiats was the aircraft used by Juan Mendoza Innernuldio, of Oruro, to make the first flight between Poopó and Oruro on 21 November 1921, his machine being described as a "Fiat con motor Ansaldo de 300hp."

The situation during this formative period is clouded further by yet another JANE'S report, which appeared in the 1928 issue of that annual. In it, the correspondent reports, without qualification, that at a Centenary Celebration held in August 1924, over 40 aircraft participated, although this report is thrown into question when it went on to state that the Cuerpo was organized into a Guerra or "war" group and an Escuela or "school" group and that the "war" group had Fokker C-Vs, Breguet XIXs and the "school" group Morane-Saulnier and Caudron C97 trainers. Of course the Fokkers and Breguets are known to have not arrived until well after that date. It can only be assumed that the 1928 issue of JANE'S somehow garbled the editing of this report.

Near the end of his tenure, and following the acquisition of both the Fokker C-Vs and Breguet XIXs, Major Haeberli was joined by the French Mission of M. Demarts, who trained Bolivian pilots, apparently, exclusively on the new, high-powered aircraft. The Fokker C-Vs, on the other hand, seem to have been introduced to the Bolivians by a little-known mission led by a European named Meinecke.

Meanwhile, in Paraguay, military aviation had only a slightly later start than in Bolivia.

In 1912, possibly following the very first flight in Paraguay on 25 November by the French aviator Marcel Paillete from Buenos Aires in a Farman biplane, Sub Teniente Silvio Pettirossi of the

Paraguayan Army was sent to France to attend the Deperdussin flying school at Reims, where he graduated the following year. Pettirossi returned to Paraguay with a new Deperdussin "T" monoplane which constituted the very first Paraguayan military aircraft. However, although accounts are conflicting, this same pilot reportedly crashed to his death in this aircraft near Buenos Aires some three years later in October 1916, bringing aviation in the country to a standstill. Tte. Pettirossi thus became the first aviation martyr in Paraguayan military aviation history.

While Pettirossi was away in France, however, the Government at home engaged the services of a young Chilean pilot, Eduardo Molina Lavín, to serve as organizer and instructor for what was termed, circa 1914, the Escuela de Aeronautica Militar. This young pilot managed to acquire, apparently from Chilean and Argentine sources, two Bleriot's with 50hp engines, a Farman also with a 50hp powerplant, and an aircraft described as an "Anzani" with a 35hp engine. The fate of this endeavor is not known, however, nor are the fates of these pioneer Paraguayan aircraft.

Sometime early in 1918, however, another Paraguayan Army officer, Arturo Escario, was also sent to Europe where he was charged with responsibility to acquire "the most modern aeroplanes for use by the Army Aviation Training School near Asunción," according to Aerial Age Weekly for 8 April 1918. The agenda (and results, if any) of Escarios mission have not been located, although Escario figured prominently in the fate of one of the very first Paraguayan military aircraft.

In August 1919, an Italian Air Mission then resident in neighboring Argentina sent a single Lohner Type L.3 and a Macchi-Nieuport M.7 to Paraguay at the request of the government to aid in promoting the acquisition of aircraft, the latter, described as "M-T-1" conducting a flight over the capital, Asunción, on 8 August. The Lohner was subsequently ceded to Paraguay on 28 August after its return to Argentina and was handed over to Paraguayan Tte. Arturo Escario over the objections of an NCO member of the Italian Mission, Sgto. Mario Conforti, who pointed out that Escario had yet to even win his pilots wings! Escario, apparently after some instruction, attempted to fly the aircraft home starting on 1 October 1919. However, following a somewhat hesitant take-off from Buenos Aires, he crashed some 20 blocks away at Playa del Pajarito. Escario died of multiple skull fractures on 2 October, thus becoming Paraguay's second known aviation martyr (after Tte. Pettirossi).

Like Bolivia, Paraguay also sent at least one officer to the Argentine aviation school at "El Palomar" for instruction. This officer, Victorio Barbero, received his brevet in 1921, although he was killed in a training flight before the Army could benefit from his skills, Paraguay's third aviation officer to die in line of duty.

Economics in the post-WW1 period precluded any serious attempts to re-kindle military aviation in Paraguay, although an initiative by the Minister of War resulted in the paper creation of the Escuela de Aviacíon Militar in 1921 at Campo Grande under the direction of Captain Hernan Dalquist and Tte 1° Oscar Grave, a German pilot with WW1 experience. However, in May 1922, a civil war saw both factions introduce aircraft into the country, flown apparently entirely by mercenary personnel. At least 12 aircraft of various types reached Paraguay including at least one Armstrong Whitworth F.K.8s (the only aircraft of this manufacturer known to have reached Latin America), one Spad-Herbemont S.20, two S.A.M.L. S.1s, five Ansaldo SVA.5s and two Ansaldo SVA.10s. These aircraft apparently arrived from both Brazil and Argentina, but it is also known that the Government forces were presented with two aircraft on 11 April 1922 by the German Government, of all places, although the types involved seem to have escaped the record. One other aircraft, an Airco (de Havilland) DH-6 was being ferried from Buenos Aires to Asunción by a British pilot, a "Mr. Brown," to be used by the Government against the rebels, but it crashed at Corrientes, Argentina enroute. Another DH-6 was known to have been used in Paraguay circa 25 March 1922 by the Centro Nacional de Aviación, although as this was apparently intended for the training of civilian pilots, it is not clear if it became involved in the revolution. Little is known about this apparently short-lived school.

The final aircraft that was to have been involved in this little-known conflict was a Caproni hydroplane which was presented to the Government by the Italians, apparently via their mission in Argentina. However, it too was crashed, reportedly under the control of a Lt. Escario, on 23 June 1922, the same day as the DH-6. Although the aircraft which actually arrived in Paraguay seem to have been utilized in a rather desultory fashion - there being little reconnaissance or intelligence regarding the disposition of opposing forces and thus few worthwhile targets - at least one of the Government aircraft, one of two Armstrong Whitworth F.K.8s, did in fact carry out a bombing attack on rebel forces in May 1922 and was reported to have been named "Presidente Ayala" (this aircraft always previously having been identified as a Breguet XIV). Flown by an English mercenary named Stewart and observer Sgto. Cusmasich, it dropped at least four bombs near Pirayú. Another attack may have been carried out by Sergeant Nicolas Bó, a former member of the Italian Mission in Argentina, who flew primarily Ansaldo SVA aircraft. This pilot stayed on in Paraguay after the civil war to instruct native Paraguayan pilots at Campo Grande until a French Mission was contracted.

Of the aircraft involved in the revolution, five survived to become the first aircraft of the formally established Paraguayan Army aviation school four years later. This Mission was, at least in part, amongst the measures secretly being undertaken by Paraguay as early as 1924 to prepare for what eventually became the Chaco War. During that year, the Paraguayan Army stood at a total of 3,000 men, all ranks. Plans were set in motion to expand the regular Army to 4,000 and to purchase modern arms and equipment, including aircraft. The utmost secrecy was exercised during this period, in order to prevent either the premature outbreak of hostilities or to provoke Bolivia into an arms race.

The French Mission itself arrived in Paraguay 31 March 1926, headed by Colonel E. M. J. Coulet and Major H. C. Luis Fromont, who was the head pilot and observer instructor. Upon the advice of the Mission, the Government contracted for three Hanriot H.D.-32E.P.-2 trainers, single examples of the Morane-Saulnier M.S.35E.p.-2 and M.S.139 and four 80hp "semiplanos intercambiales," which have remained unidentified, but which were probably "Pinguinos," the non-flyable, underpowered taxiing trainers favored by the French during and immediately after WW1 for ab initio training (although Paraguay also used at least one of the S.A.M.L. S.1s for this same purpose, minus its wings). All of these aircraft arrived in Paraguay in April 1927, where they joined the surviving aircraft of the 1922 revolution (of which, by 13 February 1925, three were still in "good condition," the other three "undergoing repair"). The first class of Paraguayan pilots graduated the following year. One of these, Tte. 2° Trifón Benitez Vera, went on to serve in combat during the Chaco War, and became one of the victims of the Bolivian "ace" Rafael Pabon.

The "Schenone Plan of 1925" (named for the French Army Mission chief, General Manlio Schenone) had outlined a Paraguayan Army aviation force of not less than 64 combat aircraft, consisting of 32 pursuit aircraft, 16 bombers and 16 reconnaissance aircraft - a number considerably beyond the means of the Paraguayan treasury at the time. Instead, funds were approved for the purchase of eight pursuit aircraft, eight bombers and six trainers - the first of which to arrive were the Hanriot and Morane-Saulnier aircraft noted earlier. The funds available were juggled to arrive at the right amount to acquire the first six Potez 25A.2s, which were delivered in 1928, and seven Wibault Type 73 C.1 pursuit aircraft.

Finally, it should also be pointed out that Paraguay engaged the services of an Argentine military mission as relations with Bolivia became ever more strained, an event that has seldom been reported. This mission included the services of aviator Jorge E. Souville, although its participation in the overall picture is not clear.

Thus it can be seen that as of late 1927, both Bolivia and Paraguay had, within their means, taken such steps as they could individually afford to avail themselves of some form of aviation forces with at least some combat capability. The stage was set for what was to follow.

1

Opening Hostilities

On 5 December 1928, an incident occurred which very nearly precipitated the eruption of full-scale hostilities four years before the final escalation (which eventually took place).

A Paraguayan Army unit based at Bahia Negra, far up the Rio Paraguay, under the command of Major Rafael Franco, attacked and, in apparent contravention of standing orders, took Bolivia's Fortín Vanguardia and Golpón. Although there is some disagreement on the actual impetus for this attack, it is generally recognized that Major Franco was acting on his own initiative at the time.

The Bolivians, initially somewhat stunned by the unexpected and unprovoked attack, responded by recapturing Vanguardia on 8 December, although an attempt to re-capture Galpón was unsuccessful. Diplomatic relations between the two nations were severed, after a flurry of diplomatic exchanges and public statements of outrage and preliminary mobilization measures of the respective armies were undertaken.

One of the earliest aeronautical events related to the Chaco region, however, involved neither a Bolivian nor Paraguayan aircraft but, rather, a French Latécoère 26 transport flown by the famous french pilot Jean Mermoz. In company with two crew members, Mermoz and his aircraft were arrested by Paraguayan Major Félix Cabrera in August 1928 while flying between Puerto Suárez and Asunción. At first thought by the press to have become prisoners of native Indians in the Chaco region as a result of a forced landing in the wild, it was subsequently suggested that Mermoz and his aircraft were on an espionage mission. The Government wisely ordered the aircraft and crew released.

Prior to this bizarre event, however, Bolivia displayed her own overt preparations for conflict when the Government took the unusual step of, essentially, impressing the entire fleet of Lloyd Aéreo Boliviano into the service of the Army. Enabled by Supreme Decree of 7 March 1928, this move was justified as in the "best interests of the defense of the country." At the time, LABs fleet consisted of six sturdy Junkers F13s, which will be discussed later in this account. It goes without saying that the assets of LAB, which continued to grow throughout the period of the conflict, served Bolivia's interests very well indeed. Paraguay, on the other hand, did not enjoy the advantages of such an efficient and well-established air transport system, and had to make do with what could be scrapped together.

Following the opening moves by Paraguay, Bolivia attacked and captured Fortines Boquerón and Mariscal Lopez in the far southwestern Chaco region, and Bolivia's Grupo de Combate had flown its first operational mission, with an air attack - believed the very first of the conflict - on the Paraguayan Army and Naval base at Bahia Negra.

This first attack, however, was not especially successful, the old bombs apparently either failing to explode or sinking too deep into the mud of the area to cause significant damage. Flown apparently by a mix of Breguet 19s and Fokker CVbs, one of the latter, crewed by Tenientes Lozada and Manchego, was brought down by ground fire and the crew captured.

To enter into a full scale war at this juncture was still obviously unwise for both countries, and the initial flurry to mobilize in both nations had collapsed in disorder. Mediation by neighboring states, principally Argentina, Brazil, Chile and Peru provided a convenient means for both parties to "save face," and a new treaty, signed on 5 January 1929, temporarily restored the peace. Subsequently, the embarrassing attempts to mobilize were discontinued and, on 22 July 1930, diplomatic relations were restored. As part of this process, Bolivia relinquished the captured fortíns.

In the interim, Bolivia's Cuerpo de Aviación commenced a determined aerial survey of the entire disputed area, using LAB Junkers F13s amongst its other reconnaissance types, and in the process had taken note of the Pitiantuta Lake region, where the Paraguayan Army had begun construction of the Fortín Carlos Antonio Lopez. For its part, Bolivia had established a forward General Headquarters, together with a depot and substantial airfield at Villa Montés, which was to prove of pivotal importance in subsequent events. Other rather crude "aerodromes", really nothing more in most cases than cleared flat areas, were established at Puerto

Suarez, Muñoz, Carainditi and Ballivían, in preparation for a renewal of hostilities.

The respite was indeed relatively short-lived. Further skirmishes and incidents occurred in September 1931, when Bolivian ground forces seized the abandoned Paraguayan Fortín Masamaklay, from which a Paraguayan force subsequently failed to evict them. Construction of more of these primitive defensive positions in the unspeakably hostile Chaco region intensified, and both sides continued to enlarge their ground forces in the area.

In June 1932, all of these moves finally brought events to the boiling point and Bolivia managed to occupy the strategically vital Fortín Carlos Antonio Lopez, which commanded the only year-round abundant water supply in the central Chaco region, a resource of incalculable importance. Paraguayan efforts to retake the position failed, but the Bolivians evacuated the Fortín, moving their forces in the area north-eastward to establish a new defensive position, Fortín Mariscal Santa Cruz, a locale better suited to their own requirements but still controlling the vital water supply.

On 11 July 1932, peace talks once again broke down into a series of accusations and counter-accusations, and Paraguay once again ordered full-scale mobilization. The battle was about to be joined in earnest.

During this period of some three and one-half years, both sides, as will be seen, had taken steps to increase and modernize their aviation forces. These measures can only be described as frantic, some successful and others not so and, to some extent, the success of these had much to do with the subsequent events as they unfolded.

At this point, this narrative will take up an examination of the aircraft involved, in whatever fashion, with both sides throughout the period 1928-1935. American aviation historians have previously recorded these in a rather arbitrary manner, based on the relative importance and activity of the aircraft in the course of the war to both sides.

2

Bombers, Reconnaissance, and Multi-Purpose Aircraft

THE CURTISS-WRIGHT C14R "BOLIVIAN OSPREY"

Designed originally by Fred Landgraf, the Curtiss-Wright CW-14 family of aircraft were developments of the Travel Air Model 4 and 4000 biplane series. Travel Air had been acquired by the Curtiss-Wright empire during the depression chaos and its well-known product line redesignated to conform to the Curtiss marketing scheme.

Initially, the popular three-seat civil and sport versions of this series carried on the old Travel Air "Speedwing" marketing designation, while the two-seat export military models were given the popular name "Osprey." This has led to some confusion amongst historians over the years as, for example, two Travel Air "Speedwing Sportsmen" sold to the Government of Panama as such in September 1931 were, for all intents and purposes, armed "Ospreys." Others, exported officially as C14Rs, were sold to the air arms of El Salvador (three, c/n 2033 to 2035 in late August 1933), Colombia (at least three as early as December 1932), Ecuador (two, c/n 2014 and 2015 in mid-August 1932) and Venezuela (probably two, c/n 2012 and 2013 in early August 1932), but the exact number of C14Rs built remains something of a mystery.

Believed to be the very first C14R "Osprey" for Bolivia in full Bolivian markings at the factory, minus serial number. Both the fixed forward gun and the rear gun were in place. Note the short tail wheel configuration, later altered in Bolivia to improve take-off and rough field handling (Hagedorn Collection).

Curtiss-Wright actually marketed two basic variants of the design. These were the C14B "Osprey" with a 300hp Wright engine, and which had the cockpits moved slightly aft and a different center section strut arrangement from the C14R. One of these, the only one known, is believed to have been exported to Venezuela along with the two C14Rs.

Bolivia itself eventually acquired a total of not fewer than 20 C14R "Ospreys," making it the single most significant user of this forgotten warplane and, at the same time, making the "Osprey" the most widely used aircraft on either side during the conflict. These aircraft were acquired in three batches: the first, consisting of 12 aircraft, had arrived by 31 January 1933 and was followed by another follow-up order for six more in May 1933, and one final order for three replacement aircraft signed in September 1933, although apparently only two of these were received, their erection at La Paz not being completed (by a Curtiss employee named Berger) until around 15 December 1933. The third aircraft in this final order was held up in Customs at the Chilean port of entry and it never reached Bolivia, and its fate is unknown. A final order for six more had been agreed upon circa 21 September 1933, but the U.S. arms embargo brought this order to a halt. It is not clear if these aircraft

OPPOSITE: Three-view drawing of the principal Bolivian air weapon, and one of the most important aircraft of the Chaco War, the Curtiss-Wright C14R "Osprey" (Curtiss).

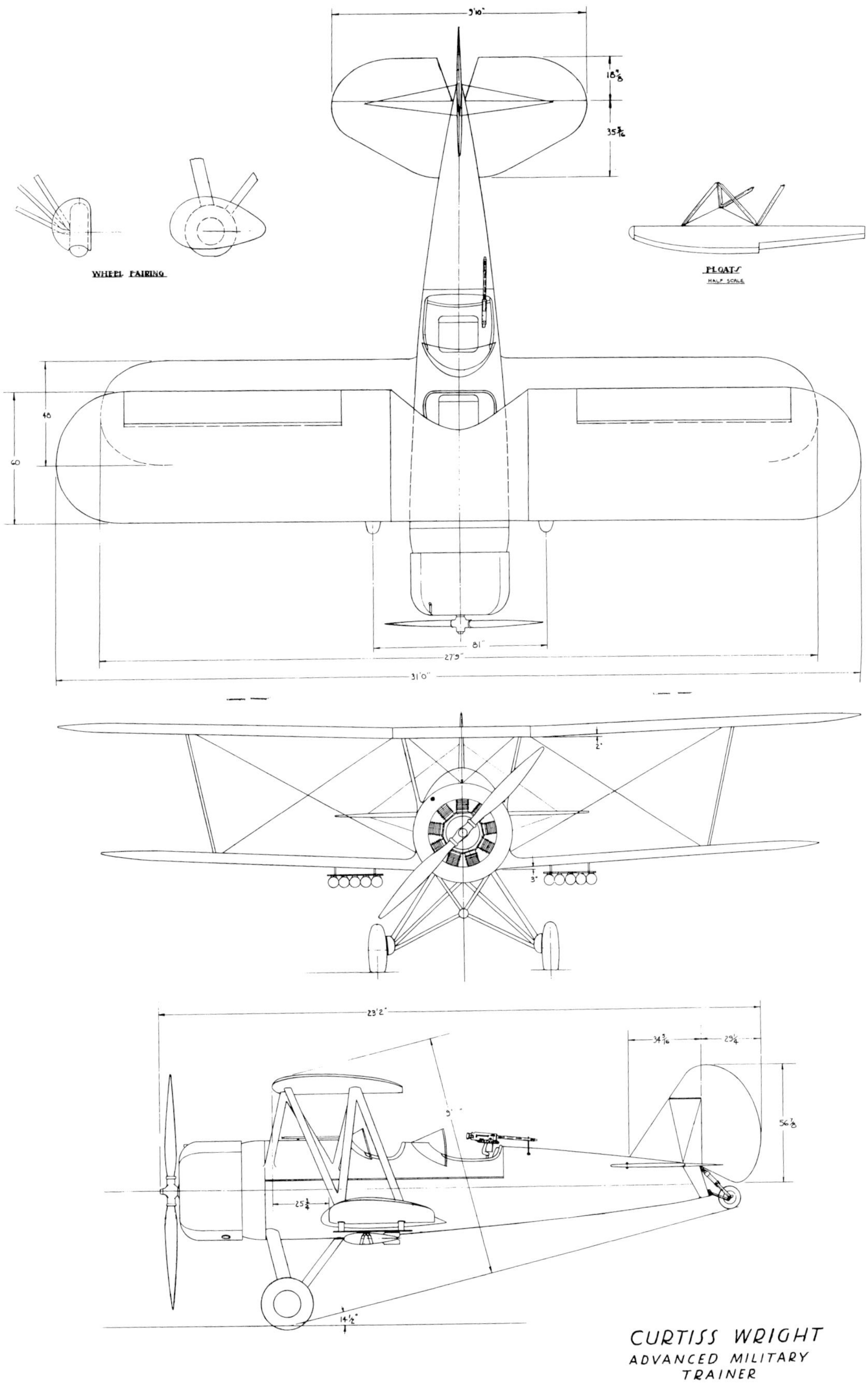
WHEEL FAIRING
FLOATS
HALF SCALE
81"
27'9"
31'0"
23'2"
CURTISS WRIGHT
ADVANCED MILITARY
TRAINER

A typical Bolivian line-up showing the average ration of Curtiss "Hawk II" fighter-bombers to "Osprey" all-purpose aircraft. Probably Grupo 2 at Villa Montes, this 1933 view possibly shows the first three "Ospreys" as none as yet have serial numbers (Ramiro Molina Alanes Collection).

were actually proceeded with in St. Louis or not and, if so, what became of them.

It is interesting to note that, on its own documents, Curtiss-Wright referred to these aircraft most often as "C14R Bolivian Ospreys," although they were also called "Curtiss Wright Advanced Military Trainers" on several occasions, and, in at least one sales document as "Curtiss-Wright C14B-9s," perhaps as a subterfuge to placate nervous export authorizing officials.

It is no overstatement to say that the "Osprey," with its 420hp Wright R-975E-2 engine, robust construction, reliability and durability, was (without question) the favorite of Bolivian airmen during the course of the Chaco War's most intense air action. Because of this, however, attrition on the type was correspondingly high, and by 24 March 1933, six of the first 12 acquired had already been lost to various causes, although five of the wrecked examples were apparently repaired (or were capable of being repaired) when the necessary spares arrived. Of the two written-off to this point, one had been shot down by ground fire and one "a complete washout when the pilot was ordered to land in a stump-infested field," according to a Bolivian report. Another "Osprey" suffered seemingly endless indignities during the first week of September 1933. A Mr. "Don" Donnelly, a skilled mechanic with the Curtiss-Wright team in Bolivia (and who was made a Captain in the Bolivian Army, at his request) was sent to the Chaco to aid in the repair of some of the damaged C14Rs. Upon his return to La Paz, he was confronted by one which (in the words of "Cliff" Travis) "one poor devil had terrible luck with. He had made three successive flights [at La Paz] with an "Osprey," and had minor crack-ups on each flight! On his fourth flight, he unwisely let the ship swerve on him on take-off and he just about had flying speed when he connected with a large rock piece at the edge of the field. It sheared off his landing gear cleanly and then rolled him end-over-end three times, but didn't hurt him a darned bit. "Don" gets $75 per week plus living expenses, but he will earn it when he repairs that wreck!" It is not clear if this aircraft was ever repaired or not, but the survival of the pilot in so spectacular an accident didn't hurt the aircraft's reputation for ruggedness a bit.

A fully marked Bolivian "Osprey," serial 91 reveals the relatively compact size of the all-purpose aircraft. The thin cheat-line down the fuselage side and engine cowl have been reported as both black and red by veterans of the conflict! Black seems more likely (via Georg von Roitberg).

It is interesting to note that Bolivian authorities took Curtiss-Wright to task for some of these losses, and put the firms representative on the spot when they alleged that the aircraft's very short tail skid had contributed to some of the training and operational losses in landing accidents. As a result, Curtiss-Wright engineers quickly designed a much higher tail-skid as well as an alternative tail wheel arrangement for the Bolivian "Ospreys" and these had reached La Paz by late March 1933. Unfortunately, all six of the surviving aircraft were off in the Chaco by that time, and retrofits to these were only accomplished piecemeal. The Curtiss engineers apparently raised quite a row about this issue. C. K. "Cliff" Travis, the C-W representative, reported from La Paz that the new skid, which could be "cut locally to any height desired, decreased the take-off run [at La Paz] by a good lot and the landing speed, if increased at all, is so little as to be unnoticed." He found that the take-off run at La Paz with the new skids was some six-seconds less than previously. The new skid also enabled pilots to "keep the rear fuselage off the ground at Villa Montes and Muñoz, where the sand is so deep."

Bolivia's aircraft purchases from Curtiss-Wright Export Corporation are a study unto themselves. Initially, the Corporation had dealt with the Government through a number of somewhat less than satisfactory agents (one of whom was F. "Pancho" Echenique of Los Cerrilos, Chile and another, Sr. Cueto Pozo, who was in fact in the employ of the Bolivian Government at the time!), paying commissions to these agents for sales generated. By 9 February 1933, the Bolivian Comptroller General, Sr. Lopez, had become "very much upset" by this state of affairs. Sr. Lopez insisted that all business with Curtiss-Wright was, henceforth, to be conducted directly between the Government and the Corporation, and that "no commissions would be permitted."

Sr. Lopez went so far as to threaten to discontinue all business relations with Curtiss-Wright if this condition was not met. C. W. Webster, President of Curtiss-Wright Export, receiving this information direct from the Government, made the sobering revelation that "in some way or another the Government, through its Intelligence Service, has learned of our dealings and prices with other countries [apparently in connection with the sale of essentially identical "Ospreys"] and what commissions were normally allowed." Lopez' warning was not entirely righteous indignation, however, as it was later learned that he was a silent partner in the firm Activos, which later assumed agency in Bolivia for the sale of Curtiss products! As Lopez, "our tall friend," was described by Cliff Travis as being "the deciding factor in all purchases - more so than the President himself," his interest in this business seems to have been more than met the eye, but Travis hastened to add that management should not "get him wrong; Lopez is one of the few honest men in the country, and although he has a personal interest in the sale of Curtiss planes and equipment, he would not let the order go through unless the stuff was entirely satisfactory."

Although "Osprey" losses during the war are fairly well documented (these are described later in this narrative), Curtiss-Wright itself took note of one of these, one of the earliest known, in connection with its "urgent" order for 18 parachutes. Shortly before 29 March 1933, one of the C14Rs operating in the Chaco "was shot down and the observer, uninjured, jumped from the falling aircraft with the intention of landing in the Rio Paraguay. He hit the river alright but, as he jumped from 1,000 feet with no chute, it was just too bad." The pilot and aircraft, and needless to say the observer, were all lost.

The "Osprey," although eventually viewed with affection by Bolivian crews, had not always been popular. Lieutenant Colonel Alfredo Santalla, for instance, viewed the relatively light-weight combat aircraft with disdain when they first arrived at La Paz. By March 1933, however, after 15 hours at the front on the type, he returned to the capital and was described by the Curtiss representative in La Paz, Cliff Travis, as "plumb goofy over our planes...and can't say enough in our favor."

Bolivia paid an average of $18,000 for each of its C14Rs (compared to $25,000 for a single-seat Curtiss "Hawk II" fighter-bomber), with 10% allowed for "commission." It is important to keep in mind that, to this point, Bolivia had not as yet acquired any Curtiss "Falcons," although by 30 May 1933, Cliff Travis had "started talking Falcon" to the authorities. The Bolivians responded that they were "quite pleased" with the "Osprey," and "can get just as much out of them for $10,000 less." Although Travis pointed out to the Government that the "Falcon" would perform much better at the altitude of La Paz than the "Osprey," even at this early juncture the General Staff envisaged establishing a permanent base for an "Osprey"-equipped unit at Villa Montes (which was at sea level) and thus the performance difference was inconsequential. Travis also reported home to Curtiss-Wright Export president Webster that "...our planes have created a real menace to the Paraguayans, as well as a deep fear. The Chief of Bolivian Air Corps operations in the Chaco has

Although more rugged than it might appear, the very primitive operating conditions in the Chaco region took a far greater toll on aircraft than the aerial opposition. This "Osprey" was probably eventually repaired, but only after enormous effort (via Georg von Roitberg).

been urging the Government, for several weeks, to buy six more "Ospreys" and three additional "Hawks", the final decision resting with none other than General Kundt, Commander in Chief of Field Operations."

Although several discussions took place with regard to modifications to Bolivian C14Rs by the manufacturer (besides the aforementioned tail-skid/wheels), including at least one lengthy exchange involving a larger fuel tank on one of the intermediate deliveries, all 20 aircraft eventually received were standard models. These were equipped to accommodate either five 30-pound, three 50-pound or two 120-pound bombs, or any combination of these totaling not more than 260 pounds. Bolivian crews usually left the rear gunner/observer behind if flying with the maximum bomb load, as performance and range suffered rather seriously if the aircraft was attempted to be operated at all-up weight and armament configurations, even at Chaco sea level altitudes. The Bolivian "Ospreys" had provision for a radio receiver, but it is not clear if these were acquired or not, although the entire aircraft was shielded and bonded for radio installation (although the engine was not, this involving an extra cost). It is not generally known that the rear gunner/observers cockpit was also fitted with an air speed indicator, altimeter, tachometer, oil pressure gauge and engine switch. Additionally, provision was made for a camera in front of the gunners seat in the rear cockpit by means of a detachable door hatch door in the bottom of the fuselage. Apparently at least a few of the aircraft were used for aerial photographic reconnaissance at one time or another.

Armament, besides the bomb loads mentioned earlier, consisted of one forward-firing synchronized machine gun mounted on the right hand side of the forward cowling, and was fired by a standard U.S. Army Air Corp type trigger. A removable charging handle for the .30 caliber weapon (7.7mm) was conveniently located for the pilot and a standard USAAC gun sight was mounted on the upper fuselage just forward of the windscreen. The gun could accommodate a maximum of 500 rounds of ammunition. The rear, flexible gun could be moved very rapidly from one side of the cockpit to the other, or locked in any one of seven positions. It could also be moved vertically upward or downward over either side of the fuselage and over, under, or around all tail surfaces, an unusual field of fire for such a gun position. Provision was made for four spare ammunition cases totaling not more than 500 rounds for the .30 caliber weapon. Seldom mentioned is the fact that the "Osprey" was offered with both streamlined wheel "spats" similar to those worn on the "Hawk II's" as well as fittings for pontoons, although Bolivia appears not to have taken up either option.

The Bolivian "Ospreys" had a quite respectable performance for a light combat type. Maximum speed (on a really good day) was in the vicinity of 163mph, while it would normally cruise at around 142mph. Interestingly, its absolute ceiling was rated at 21,000 feet. Maximum cruising range was normally not more than 445 miles at an all-up weight (less fuel) of 2,782 pounds. Its time-to-altitude at the normal low altitudes at which operated in the Chaco was quite respectable, taking 2.7 minutes to attain 3,280 feet (1,000 meters).

All Bolivian "Ospreys" were delivered doped aluminum overall, with the Bolivian national insignia roundel in four wing positions and on either side of the rear fuselage. The national colors were also carried on the rudder of all aircraft and, somewhat unusually, the upper (red) stripe carried completely forward to the tip of the rudder. A few of the aircraft had a thin dark blue or black "cheatline" running most of the length of the fuselage aft from the engine firewall area, and all had the words "Curtiss-Wright" over "Osprey" on the forward fuselage in small, about three-inch letters on either side. The word "Osprey" was sometimes followed by the individual aircraft serial number. These numbers, which were carried in about 25 inch block black numerals on the vertical fin on either side, were 11, 13, 20, 22, 25, 30, 39, 45, 50, 60, 62, 64, 68, 78, 87, 91, 93, 95 and two aircraft which apparently never gained a serial prior to their loss. The reasons for the assignment of this rather wide and random spread of serial numbers seems to have been an attempt to veil the actual number of such aircraft of this type in use by Bolivia from Paraguayan intelligence. The only known manufacturers serial number for a Bolivian "Osprey" is c/n 2036, apparently one of the second batch, which was tested in May 1933. By way of "special" markings, at least six of the "Ospreys" are known to have had painted engine cowlings at one time, and these have been variously described as all black, all red or both colors separated in equal segments by a thin white (or possibly bare metal) line. Photo evidence seems to support the latter.

By the end of 1932, Bolivia's primary combatant aircraft in the area of operations were its few remaining Vickers Type 143

"Scout" pursuit and Vickers Type 149 "Vespa III" recon-bombers, which were getting to be rather long in the tooth, although the combined total of 12 aircraft were all, remarkably, still operational. With the apparent arrival of the first of the "Ospreys" in December 1932, however, and their rapid introduction into service, the British aircraft were at last supplemented, and the first known "Osprey" operations occurred 2 January 1933 during the first Bolivian attack on Fortín Nanawa. Flying from their base at Muñoz, some 80km distant, the "Ospreys," aided by the first of the Curtiss "Hawk IIs" and the Vickers aircraft, provided invaluable ground-attack aid to the Bolivian ground forces efforts to regain lost territory and cut the vital Nanawa-Concepción supply road.

Nanawa itself was, by the standards of the Fortíns of the Chaco region, a particularly heavily defended Paraguayan position, with a well-engineered and rather elaborate system of casemated trenches and barbed-wire entanglements. Bolivian air attacks, led mainly by Lieutenant Colonel Bernardino Bilbao Rioja, engaged in almost non-stop low-level strafing attacks in the face of very heavy ground fire. Sporadic action continued around Nanawa into late June 1933.

Meanwhile, Colonel Rioja and most of his mixed trio of escuadrillas (each nominally named at the time for the type of aircraft employed such as "Hawk," "Osprey," "Vickers" etc) were moved enmasse by General Kundt to support the attack in the area of what became known as the battle of Toledo between 11 February and 12 March 1933. There, on 25 February, in the course of executing one of the many extremely low-altitude strafing and gun attacks on the Paraguayan defenders, Bolivian Tte. Arturo Valle Peralta and his gunner, Sub.Tte. José Ardiles, were shot down in their "Osprey," both perishing and the aircraft lost. This crew thus became the first casualties due to enemy action during the Chaco campaign, and may have been the crew involved in the previously cited incident in which the observer jumped from the burning aircraft without benefit of a parachute. Another "Osprey" was reportedly shot down by Paraguayan ground fire near Caraya, although the date of this loss is not known.

At about this juncture, the Cuerpo de Aviación graduated its first wartime class of new pilots, who were badly needed at the front. Unfortunately, their inexperience soon became apparent when one of these, Sub.Tte. Mario Calvo crashed one of the "Ospreys" at Fortín Florida, this almost certainly being the aircraft cited by Cliff Travis as having landed, "on orders" in a tree-stub covered airstrip. Calvo's observer, Captain Rafael Pabón, was probably the officer who gave this order and, although both survived, the virtually brand-new "Osprey" was lost. These first two losses may very well have been the "Ospreys" that did not receive individual aircraft serial numbers.

As soon as sufficient aircraft and crews had become available, Tcnl. Bilbao, by then the Officer Commanding the Bolivian Cuerpo de Aviación, dispatched a detachment of "Ospreys" under Captain Leónidas Rojas to Puerto Suárez, from whence they carried out bombing and attack missions against the important Paraguayan river port of Bahia Negra to the south some 100km. These attacks were generally effective, one of the Paraguayan casualties on the ground being none other than the garrison commander, Colonel Julian Sánchez.

On 27 April 1933, at 1050Hrs, a flight of six "Ospreys", the largest concentration of one type thus far launched during the war, carried out an attack on Puerto Casado from their base at Muñoz, the results being described as "very effective." Unfortunately, the attack on this city and civilian targets, resulted in a strong protest from the Argentine Government, which was sympathetic to Paraguay prior to 1937, and as a result Bolivia reluctantly suspended any further air attacks on strategic targets on the Rio Paraguay for the time being.

At about this same juncture, Captain José P. Coello, a veteran flyer and formerly a Vickers "Scout" pursuit pilot, conducted a solo reconnaissance flight in an "Osprey" to the Paraguayan city of Concepción during which he apparently dropped propaganda leaflets of his own design, resulting in yet another protest from the Argentine Government. Although the Bolivian High Command denied any knowledge of this mission, it is hard to believe it had not been authorized.

In the meanwhile, Paraguay had reinforced the defenses of Nanawa, including the entire new Third Corps (consisting of the 4th and 5th Infantry and 2nd Cavalry Divisions, to which were added the veterans of Fernández's veteran 1st Division). Informed of this ominous build-up by Bolivian air reconnaissance, General Kundt shifted his attentions to the strategically insignificant Paraguayan Fortín Fernández, against which he threw an entire Division in repeated (and very costly) frontal attacks. At this point, after 10 months of undeclared, full-scale hostilities, Paraguay formally declared War on Bolivia on 10 May 1933. One of the primary reasons for this gesture, largely unheralded, was to hopefully move Chile (as well as Peru and Brazil) to a position of strict neutrality, since it was by then well known that Bolivia was receiving most of its arms and aircraft through the Chilean ports. As it developed, however, Chile exercised a rather curious and liberal interpretation of her "neutrality" towards her old enemy, Bolivia, although in fairness, it must be observed that Argentina was equally easy on her neighbor Paraguay.

Kundt then wisely redirected his efforts towards Nanawa, leading to what the Bolivians call the second battle of Nanawa, commencing on 4 July 1933. Operating once again from Muñoz, "Ospreys" and "Hawk IIs" were specifically tasked with attacking Paraguayan artillery positions and aviation assets (although these were very elusive and few targets availed themselves). By mid-July, however, after the loss of at least 2,000 ground troops in the face of the determined Paraguayan defenders, the front settled down to static warfare for the remainder of the year until December.

The lull proved costly to the Bolivians, however, as a renewed Paraguayan offensive commenced on 23 October 1933 and pressure along virtually the entire front forced the Bolivian Army to yield ground, in spite of dogged resistance. Bolivian strength on the ground and, to a certain extent, in the air, was even more hard-pressed than usual, since General Kundt had granted leave (and relief from the tortures of the Chaco region) to large numbers of personnel, especially officers.

Despite very good aerial reconnaissance and warnings of Paraguayan movements on the ground by, amongst others, "Osprey" crews (including Captain Leónides Rojas, Juan Antonio Rivera and José Arzabe and their observers, Captain Claudio Moreno, Alfredo Pacheco and Tte. Alberto Paz Soldán), General Kundt belatedly ordered a retreat, but not before the entire Bolivian 4th and 9th Divisions were forced to surrender at Campo Via on 11 December due to a complete lack of water. This was unquestionably the greatest victory for Paraguay (or either side) of the war, resulting in the loss to Bolivia of over 8,000 veteran troops with their personal arms, at least 500 automatic weapons, 25 mortars of various caliber, 80 vital vehicles and 20 artillery pieces. This debacle proved the undoing of General Kundt, who as recently as September, had been feted in La Paz as the Bolivian's "Teutonic messiah" in the words of one historian. Indeed, the heroic actions of the Cuerpo de Aviación during this defeat were amongst the only bright spots for Bolivia, the aircraft of the small forces at hand aiding in the escape of about 2,500 men under Colonel Enrique Peñaranda from Campo Via, who proceeded to Saavedra, where new defenses were hastily organized. Colonel Peñaranda was promoted to flag rank and replaced General Kundt as Commander in the field.

Fortunately for Bolivia, the Paraguayan Army was unable to bring the stunning victory at Campo Via to its logical conclusion, e.g. the complete destruction of the Bolivian Army, as Paraguayan casualties had also been very high.

Following this great battle, new diplomatic initiatives and pressure by the "Big Three" Latin American neutrals and the League of Nations once again succeeded in gaining a truce, which took effect at midnight on 9 December 1933. The Paraguayan leader in the field, however, Major General Estigarribia, intent on capturing the key Bolivian position at Muñoz prior to the commencement of the truce, launched his exhausted troops towards that point, where he in fact prevailed, although arguably some time after the truce was supposed to go into effect!

231. Avión de bombardeo y observación, alistándose

An operational Bolivian "Osprey" somewhere in the Chaco being "bombed-up" by ground crews. When all four under-wing racks were mounted, the most common configuration during the Chaco War, the rear-gunner had to stay behind (Archivo Cnel. Av. (S.R). A. Pasmor).

As diplomatic moves once again resumed, Bolivia was in dire straits indeed. With but about 7,000 effectives remaining in the Chaco area of operations (another 8,000 were in what was called the "communications zone" between Villa Montes and the closest Bolivian transportation point), she had lost a staggering 14,000 killed, with another 32,000 having been evacuated as a result of wounds or swamp induced illness. Not less than 10,000 had been captured at various points and an estimated 6,000 had deserted into Argentina, rather than face what seemed certain death in the merciless Chaco.

The Chaco truce was short-lived however, ending on 6 January 1934.

"Osprey" strength was, as noted earlier, replenished in the meantime, however, and while frequent reconnaissance missions were flown between the time of the fall of Campo Via and the end of the truce period, the next action in which the type was engaged did not occur until 19 to 24 May 1934 during the fight for the oddly named Cañada Strongest. Reconnaissance flights over the Basguensen road from the new operating base at Ballivián, especially, provided the Bolivian command with vital intelligence on the movements of Paraguayan forces, and "Osprey" aircraft flew the bulk of these solitary and dangerous flights. Perhaps more correctly called Cañada Esperanza, the Bolivian victory here was a boost to the morale of the nation, and indeed was the greatest Bolivian land victory of the war. Some 1,500 Paraguayans were surrounded there and surrendered, although several other regiments slipped through Bolivian lines and escaped. An "Osprey," s/n 30, was lost on 22 May 1934 during this campaign, killing pilot Captain Alfredo Pascoe, although this was officially listed as a "training accident." His gunner survived.

Next scene of "Osprey" involvement was at the battle for Condado between 18 June and 8 July 1934, part of the Bolivian defense of the key position at Ballivián. With Paraguay intent on turning the Bolivian left flank, the deployment of virtually every airworthy "Osprey" to Cururenda (under Major José P. Coelle, and also including several "Hawk IIs") enabled the aircraft to shuttle missions to the scene of fighting at El Condado, some 30km distant.

It was during this defense that Bolivian "Ospreys" saw the greatest aerial action of the type during the entire war. In one action, in which both a "Hawk II" and an "Osprey" flown by Captain Rivera were apparently involved, the Paraguayan Potez 25 piloted by Arsenio Vaesken was hit repeatedly, while return fire from the Potez holed both the Bolivian aircraft as well. However, this unique aerial

One of the largest known concentrations of C14R's, this lineup of six airworthy examples, probably at Villa Montes, can be identified as a very early, post-delivery image, as none of the aircraft yet have serial numbers on their vertical tails (via Antonio Sapienza).

combat proved inconclusive and all three aircraft survived to fight another day.

On 12 August 1934, however, an "Osprey" flown by Major Rafael Pabón and his gunner Suboficial Mario Calvo, while conducting an armed reconnaissance in the vicinity of Fortín Florida before the 3rd Corps area encountered another Paraguayan Potez 25 and, after repeated passes with his more maneuverable aircraft using the forward gun, allegedly shot the Potez down although his aircraft, s/n 78, was also lost in the contest. In fact, this Bolivian claim was in error, as the Paraguayan Potez involved, No.11, returned to base safely.

While Bolivian pilots were distinguishing themselves, their forces on the ground were being outmaneuvered and humiliated by the skillful Paraguayan leadership. Bolivian "Osprey" aircraft continued to operate until the end, however, including the battles for Isopoirenda and Algodonal between 6 and 24 September 1934, the now designated Grupo Aeréo being consolidated and operating exclusively from the beleaguered Bolivian bastion at Villa Montes. A small force of "Hawks" and "Ospreys," however, designated as the Grupo Aeréo Santalla, operated intensively against advancing Paraguayan forces from Carandaití, flying distances of up to 150km from their base in support of beleaguered Bolivian ground forces. On at least one occasion, an "Osprey" flown by Tte. Alberto Paz Soldán and his gunner Sub.Tte. Sinecio Moreno encountered patrolling Paraguayan Fiat CR.20 fighters over the front, and, while the Bolivian aircraft was holed in the 23 September action, one of the CR.20s was reported to have failed to regain its base, and must thus be considered at least a "probable" by this most unlikely victor.

Bolivian ground forces managed to retreat in some order from the September battles near Villazón, and the next major engagement did not develop until 11 November 1934, when the Paraguayan First Corps commenced its encirclement of the Bolivian 10th Division at El Carmen, where it was shortly joined by the reinforcing

either five 30-lb., three 50-lb. or two 120-lb. bombs, or any combination totaling not more than 250 lbs.

One signal pistol holster and shell receiver can also be furnished.

Radio

Space is provided for a radio receiving set and the entire airplane is shielded and bonded for radio installation. Engine shielding is optional, furnished at extra cost.

Electrical Equipment

1. There is provision made for the installation of an electric generator.

and green 11th Division. Lack of Paraguayan air reconnaissance, however, is blamed for their failure to appreciate the size of the encircled forces. In spite of this, and totally encircled, the Bolivians surrendered after several half-hearted attempts to break out. On 17 November, the Paraguayans entered and took Ballivián, which had come to stand as a symbol of the Bolivian military presence in the Chaco. A Bolivian "Osprey" crashed on landing around 6 November 1934 and was written off. An "Osprey" was also used at about this point in time as a "forward air controller" for Junkers K 43 level bombers, flying out ahead of the German aircraft and spotting targets for them. It is believed that this aircraft was assigned to the famous Punta de Alas bomb unit at the time. Another "Osprey" was lost on 26 November 1934 at Samayhuate following the hasty flight from Carandaití, which was shortly overrun. Yet another, unidentified "Osprey," in a flight of three aircraft forced to alight due to zero-zero weather conditions, crashed in the wild when the weather cleared and a take-off was attempted.

By now, the Bolivian leadership were at each others throats and President Salamanca resolved to replace the entire Bolivian High Command in the Chaco. Proposing General Lanza as C-in-C, General Peñaranda, however, by diverting troops from the front, staged a coup d'etat at Villa Montes, where the President had arrived somewhat inopportunely aboard a Junkers Ju 52/3m.

Tejada Sorzano was installed as the new Bolivian President but, after yet another Bolivian defeat, the new President ordered a general mobilization to save the territorial integrity of Bolivia itself. Riding a seemingly unstoppable wave of victories, the Paraguayan Army was set to invade Bolivia proper and attempt to capture either Villa Montes or the Santa Cruz oil fields themselves, possibly both.

Deciding that Villa Montes had to be defended, the new Bolivian Commander in the field, Bilbao Rioja, who happened to be for all practical purposes the only gifted officer in the entire Bolivian High Command, directed the Bolivians to settle in behind elaborate field works at Villa Montes and defend that point at all costs. Unfortunately, Paraguayan forces broke through the central sector on 28 January 1935 and cut the only road between Villa Montes and Santa Cruz. At this point, the hitherto brilliant Paraguayan Commander, General Estigarribia, embarked on an ill advised frontal assault on the Bolivian positions at Villa Montes, which was thrown back with very heavy losses. Sorties by the Bolivian defenders at Villa Montes caused the Paraguayans to pull back a previously successful thrust into the Andean foothills to Charagua. The nearly exhausted combatants staggered back and forth several more times at odd places along the front and slowly a virtual stalemate developed. The Paraguayans, nearly bankrupt and with their very last reserves of manpower exhausted, could no longer even invest Villa Montes while the Bolivians, with the savage Chaco still fresh in their minds, realized they could never totally defeat the Paraguayans in their native lowlands. With this dizzy assessment at hand, both sides at last invited the peace overtures of the non-belligerents. The Bolivian air base at Villa Montes had, for all intents and purposes, by this time become the primary operating base of the Cuerpo de Aviación, and besides operational aircraft (which as of 24 January 1935 consisted of 11 aircraft at Villa Montes: three Junkers K 43s, two Curtiss "Falcons," two "Ospreys," three "Hawks" and a solitary veteran Vickers Type 143 "Scout"), also witnessed considerable training activities. It was at this point that it became the Bolivian practice to send off a "dawn patrol" of the defensive perimeter each morning, usually consisting of a single "Osprey" and two "Hawks."

OPPOSITE: The trusty Wright R-975E-2 "Whirlwind" engine of 420hp (as advertised) which powered the Bolivian "Osprey's." Note the cowl attachment support members, which resulted in the segmented appearance of "Osprey" cowls noted in many photos. (Wright).

The Peace Protocol, signed on 12 June 1935, went into force at 1200Hrs 14 June 1935. The Chaco War was over.

As of the end of hostilities, it is difficult to establish exactly how many "Ospreys" survived in an airworthy condition. Although almost certainly still the most numerous type in service, the remaining approximately three airworthy examples were apparently dispersed amongst three units. These were known as:

Escuadrilla "Osprey" at Ingavi/Puerto Suarez
Grupo Aereo Mixto at Cuevo and Charagua
Grupo Aereo Mixto at Villa Montes and Yacuiba

The last known "Osprey" loss took place on All Fools Day 1935, although the aircraft was deemed "repairable" if the necessary parts could be located.

In summary, at least 12 of the total of 20 "Ospreys" received are known to have been lost during the course of the war. Three are known to have survived at the time of the Armistice, and it must thus be assumed that the other five were lost as a result of training accidents that have remained unreported.

This summary is supported by the post-Chaco War Intelligence Reports available. As of 25 January 1940, two "Ospreys" were still counted on the strength of the Bolivian air arm. Interestly, the Bolivian Government made a formal request on 26 April 1945 to obtain "spares and materials to repair four Ospreys" through unused Lend-Lease funds still on their account at that time. Lend-Lease officials apparently did not approve this request. However, by 10 April 1946, the Air Order of Battle for Bolivia showed three such aircraft, all assigned at the time to 2° Escuadron de Entrenamiento at Cochabamba. The Ministry of Defense advanced a plan to turn over 17 "obsolete" aircraft to the civil flying schools circa 5 September 1944, including three "Ospreys," but the AOB for 18 December 1944 still showed all three on FAB strength, although all were unserviceable.

Incredibly, the AOB for 30 June 1954 still showed a single "Osprey" on FAB strength, although "inoperable," certainly the last of its kind on earth, and a true veteran. The fate of this historic aircraft is unknown.

A truly extraordinary photo, this Bolivian "Osprey" was actually in process of attacking Paraguayan troop concentrations somewhere in the Chaco in 1934 when photographed. (Museo de Historia Militar via A. L. Sapienza).

It may be truly said that Bolivia got her moneys worth from the Curtiss-Wright C14R "Osprey."

For the record, Bolivia officially reflects the following pilots as having flown "Ospreys" during the conflict:

Maj. Jorge Jordán
Cap. José P. Coello
Cap. Rafael Pabón
Cap. Luis Ernst
Cap. Leónidas Rojas
Cap. Alfredo Pascoe
Cap. Eliodoro Nery
Cap. Luis Garcia
Cap. Luis Paravicini
Cap. Juan Antonio Rivera
Tte. Alberto Paz Soldán
Tte. Arturo Valle
Tte. José Arzabe
Tte. Aurelio Roca Lladó
Tte. Claudio López
Tte. José Abdón Vargas Soto
Tte. Alberto Montaño
Tte. Juan Pando
Tte. Luis Paredes
Tte. José Chacón
Tte. Sinecio Moreno
Tte. Jorge Eulert
Tte. Luis Soria
Tte. Fabián Monasterios
Tte. Raúl Ernst

This discussion of the "Osprey" cannot end, however, without mention of the fact that Paraguay had also been offered "Ospreys" by the Curtiss-Wright Export Corporation at some point, but the Paraguayans, for reasons unknown, opted instead for the Fiat CR.20bis which, in the words of the official Paraguayan history of the air war, "was decidedly inferior to the Osprey." No further record of this fascinating offer has been found.

POTEZ 25 A-2 AND POTEZ 25T.O.E.

First revealed to the public at the Salon de Aeronautique in Paris in 1924, the basic Potez XXV (Roman numerals were used within the French aircraft industry until sometime around January 1928, when virtually all presentations changed to Arabic numerals) was one of the most produced aircraft of the between-the-wars period. Nearly 4,000 examples were built either in France or under license, and at least 90 sub-variants are known.

Besides the French Air Force, which used the type widely, Potez 25 variants saw service in 21 countries, certainly a record of some sort for a Golden Age aircraft type. The many variants were achieved mainly due to a wide variety of engine types used, attesting to the durability of the basic design.

The aircraft acquired by Paraguay were contemporaries of the Fokker C.Vb and Breguet 19 aircraft ordered by Bolivia, and thus were at something of a disadvantage during the course of the Chaco War, as they instead had to confront the "next generation" Curtiss "Hawk IIs", Curtiss-Wright C14R "Ospreys" and other more modern types.

In its literature, Potez described the 25 A-2 as a "two-seat observation, long-distance reconnaissance, light bombing or fighting biplane," which of course was in keeping with the between-the-wars French philosophy of a true multi-purpose type. To its credit, at least with regard to the Chaco War, the Potez 25 was indeed thrust into all of these roles and, in spite of its semi-obsolescence,

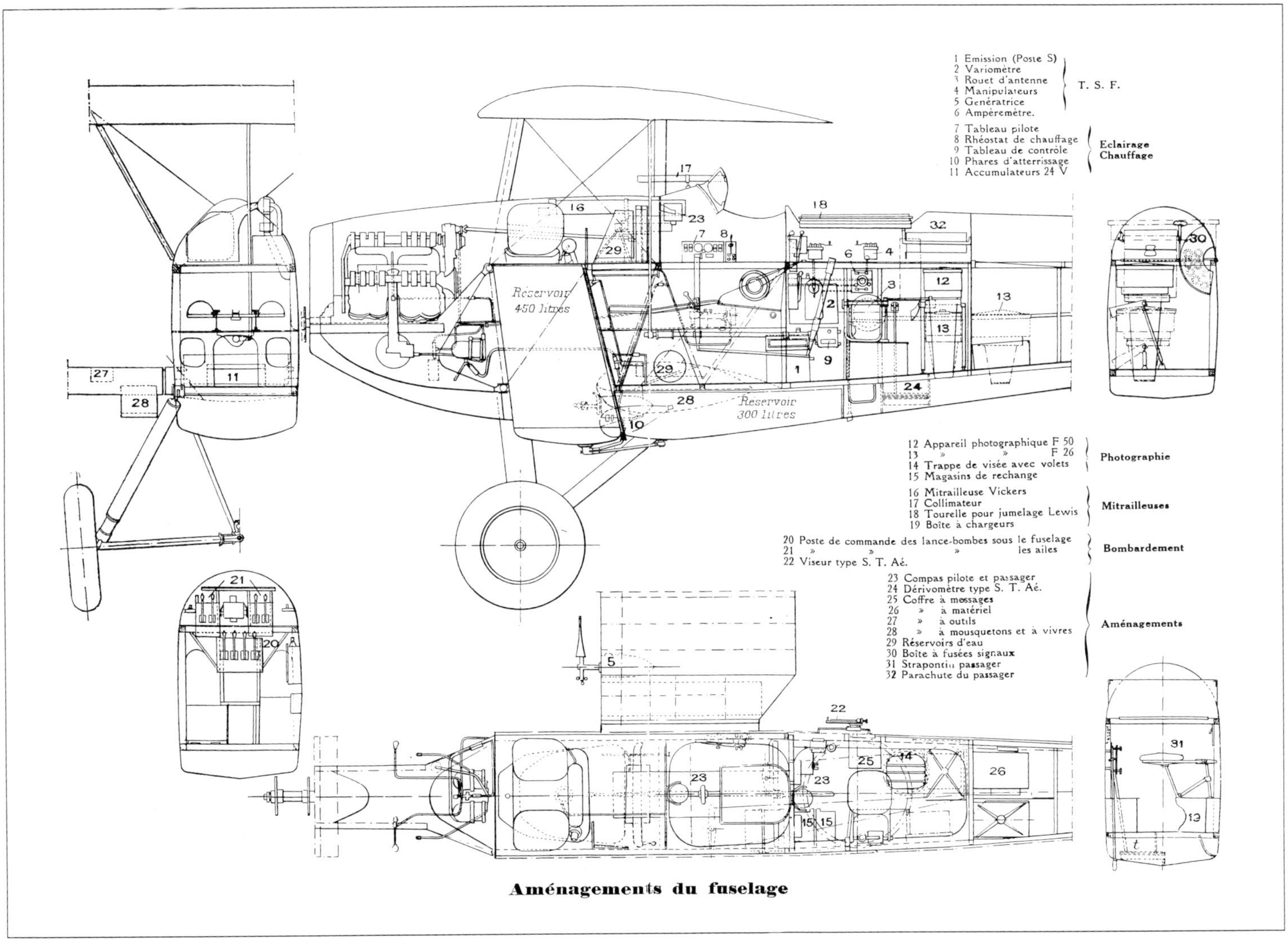

Cut-away drawing of the fuselage of the principal Paraguayan combatant aircraft, the Potez 25 T.O.E. showing the locations and identity of the major items of equipment (Potez).

Good detail view of Paraguayan Potez 25 A.2 serial 2 at Ñu-Guazu aerodrome in 1929. This was the first Paraguayan Potez 25 lost to an accident. (Archivo Cnel. Av. (S.R). a. Pasmor).

committed itself well. The Potez 25 was unquestionably the best fighting aircraft available to Paraguay's Aviación en Campaña (often cited in error as "Fuerzas Aéreas Nacionales.")

The basic Potez 25 airframe, as noted, could be equipped with any number of engines. For commonality purposes, however, Paraguay elected to order initially seven Potez 25 A-2s with 450HP Lorraine-Dietrich 12Eb engines, a tried and true French power plant which was also specified in their Wibault pursuit aircraft ordered concurrently.

Part of the overall plan laid down by the French Mission, the seven Potez 25 A-2s were ordered on 10 January 1928 at a cost of some 1,482,326 French Francs, delivery being stipulated not later than 22 March 1929. In fact, Potez managed to beat that target, the first two aircraft arriving in their crates at Asunción on 9 October 1928, apparently as a result of some production line juggling at the factory. It should be noted at this juncture that some sources have identified the first 6/7 Potez 25s acquired by Paraguay as Potez 25T.O.E.s rather than Potez 25 A-2s. This is incorrect, as will be noted further on in this narrative, when the acquisition of Potez 25T.O.E.s is discussed.

The aircraft were trans-shipped to Paraguay via Montevideo, Uruguay and it was there that one of the remaining five aircraft of the original contract was lost when, during unloading, the crate was accidentally dropped overboard into the bay. Although replaced by the manufacturer after lengthy rather heated exchanges, the badly needed replacement did not finally arrive until 1 June 1932.

The first six aircraft received thus gained serials "1" to "6" and, as soon as they were erected at Campo Grande, the only operational Paraguayan military aerodrome at the time, were immediately placed into intensive service training bomber pilots and gunner/observers. Between 1928 and 1931, a total of 25 pilots and 18 aviation mechanics were trained, although this intensive period resulted in the loss of one of the Potez 25 A-2s in 1929 at Campo Grande, but both crew survived.

It is interesting to compare the capabilities and performance of the Potez 25 A-2 with the aircraft they were to oppose in the coming months and years. With the Lorraine-Dietrich engine, maximum speed was rated at 138mph at 6,560 feet. Like the Curtiss-Wright C14R "Osprey" used by Bolivia, and against which it was often pitted in very similar usage, the Potez 25 A-2 had easily removable dual controls in the rear cockpit. Forward firing armament could consist of either one or two .30 caliber (7.7mm) weapons with 500 rounds per gun, while the rear gunner/observer could accommodate either one or two .30 caliber weapons as well. However, the Potez also had a lower fuselage trap door for another gun, although there is no evidence that this feature was ever used in Paraguayan service. For light bombing missions, the aircraft was unique in having a small internal bomb rack (described in some documents as a "Esnault-Pelterier pattern") for up to 12 light 22 pound (10kg.) bombs or two external bomb racks mounted under the bottom main planes for 16 to 24 bombs of the same size. The Paraguayan Potez aircraft appear to have been fitted with both forward firing guns and two guns in the rear cockpit, which gave the aircraft some advantages in terms of fire-power over, for example, the lighter armed "Ospreys" operated by Bolivia. The Paraguayans selected modified 1927 Madsen 7.65mm machine guns for use on their aircraft. Some of the Paraguayan Potez aircraft were also fitted with German Telefunken wireless radio sets.

With these aircraft, the Primera Escuadrilla de Reconocimiento y Bombardeo was eventually formed. In the meantime, of course, the evolving dispute between Bolivia and Paraguay for the Chaco region was becoming ever more tense. In spite of the many Bolivian incursions and erection of the many small fortified positions, Paraguay, as late as June 1932, when all-out war finally erupted, had not constructed a single landing field in the entire disputed Chaco area. The first of these, named Isla Taguató, when finally

An unidentified Paraguayan Potez 25 at the Paraguayan Isla Poí aerodrome, showing the extremely rugged conditions there and also the rarely seen auxiliary fuel tanks under the upper inboard wing panels (Archivo Cnel. Av. (S.R). A. Pasmor).

Virtually the entire Paraguayan maintenance organization at Isla Poí aerodrome circa 1932 with the (then) Lieutenant Pasmor second left. These men performed prodigious feats of engineering under the most appalling conditions (Archivo Cnel. Av. (S.R). A. Pasmor).

commissioned, was some 150 kilometers from the fighting front but, in spite of this Potez 25 A-2s s/n 4, 5 and 6 were immediately deployed there (in company with two Wibault 73 C.1 fighters), together with their ground crews. Under the command of Tte. 1° Atilio Ibanez Rojas, these aircraft arrived on 4 August 1932, but each sustained damage upon landing due to the extremely crude condition of the airfield. Reconnaissance flights commenced after some 15 days of repairs, the Potez usually being escorted by at least one of the Wibaults in the execution of these missions.

The Headquarters of the Paraguayan Army ground forces had been established at the fortress of Isla Poí (which translates to "Small Island"), and a second aerodrome was hastily scratched out there. In order to serve the Headquarters better, the three Potez 25s and the Wibault 73s were repositioned from Isla Taguató to Isla Poí, which was much closer to the front.

In late August 1932, the small operational detachment at Isla Poí was reinforced by Potez 25 A-2s s/n 1 and 3, and by a third Wibault 73 C-1, with the result that the bulk of the Aviación en Campaña, totaling eight aircraft, were now at the "front."

Action was not long in coming. During the famous battle for Boquerón, the first full-scale ground action of the Chaco War, between 9 and 29 September 1932, the Paraguayan Potez 25s and accompanying Wibaults conducted 12 bombing missions and 27 reconnaissance flights over the fortress. This resulted in the first air combat of the war, between two most unlikely antagonists. These were a Paraguayan Potez 25 A-2 piloted by Tte. Emilio Rocholl and a Bolivian Vickers Type 149 "Vespa III". The Paraguayan aircraft seems to have come out the worse, as Tte. Rocholl landed at Isla Taguató wounded, although his aircraft was apparently repaired and returned quickly to service. On 29 September, Paraguayan troops finally took the fortress.

In October 1932, another aerial combat occurred, again, between two very odd combatant types. Two Bolivian Breguet 19s attacked two Paraguayan Potez 25 A-2s which were engaged at the time in a bombing attack on the Bolivian fortress at Arce. Both Potez aircraft turned into their attackers and, after some exchanges of machine gun fire (apparently rather ineffective) the Bolivian

Breguets withdrew. The contested Fortín at Arce was taken by Paraguayan troops on 23 October.

The next target for Paraguayan forces was the Bolivian position at Saavedra, which was also the subject of numerous Potez 25 reconnaissance flights and bombing attacks. During one of the recon missions, s/n 6, piloted by 1° Tte. Trifón Benítez Vera and Cpt. Ramón Avalos Sánchez (observer/gunner) was shot down and destroyed by a Vickers Type 143 "Bolivian Scout" flown by Bolivian "ace" Cpt. Rafael Pabón on 4 December. The Potez 25 crashed southeast of Saavedra and both crew members died instantly. This was the first confirmed aerial victory of the war.

Paraguayan mechanics and other ground personnel pose before a Potez 25 T.O.E. at Isla Poí in 1933. The exhaust under the port nose one man is leaning against reportedly caused a very distinctive sound when in flight (Archivo Cnel. Av. (S.R). A. Pasmor)

In the meantime, Paraguayan officials had been vigorously attempting to acquire additional Potez 25s. By now, the obvious source, France, had become sensitive to the warnings of neutrals and the League of Nations, who were attempting to mediate the conflict, and did not wish to be seen as providing arms to either side. As a result, through the agency of the Argentine Military Mission in France, five Potez 25T.O.E.s (the odd type designation meaning Theatres d' Operatins Exterieures in French) were purchased, ostensibly for the Argentine Army, as French records simply state that the aircraft were "found abroad." The U.S. Military Attache in Buenos Aires noted these five aircraft passing through Montevideo but, incredibly, once again one of the aircraft was heavily damaged in the transshipment and had to be returned to France for major repairs. The aircraft which actually arrived gained serials 7 to 10 and were apparently followed by the two repaired aircraft and two others (presumed type 25T.O.E. and the one replaced 25 A-2 from the first contract) to make a total of 14 Potez 25s received, the final four aircraft becoming s/n 11 to 14. Some historians have been confused by the fact that s/n 8 was rebuilt following an accident and re-serialed as No.15.

Mechanics perform an engine change on a Potez 25 "in the field" at the Isla Poí aerodrome in 1933. The officer on the right is Tte. A. Pasmor, who was maintenance chief of the entire Paraguayan air arm at the time. The Potez 25 appears to have three colored bands on the lower left main plane. (Museo de Historia Militar via A. L. Sapienza)

The newly arrived Potez 25T.O.E.s, which had a rather thicker profile than the earlier Potez 25 A-2s due to the larger fuel tanks, were baptized as "Panzones" (literally, "big bellies," as the extra fuel was carried in tanks in the lower fuselage rather than in the wings as on the Potez 25 A-2s) by the Paraguayan ground crews. The first of the new 25T.O.E.s were deployed to the Chaco region commencing in March 1933, and formed the basis for the creation of the 2do Escuadron de Reconicimiento y Bombardeo.

Earlier, however, in January 1933, Bolivian forces had surrounded the Paraguayan Fortín Nanawa in a counter-offensive and, as the garrison there was completely isolated, three of the Potez 25 A-2s (s/n 3, 4 and 5) and a single Potez 25T.O.E. (s/n 7) were stripped of all but one forward-firing gun and were used as cargo aircraft in order to deliver vital ammunition to the defenders of Nanawa. The supplies were stacked in the empty gunners cockpit, something near 115 kilos of ammunition eventually being thus carried by each aircraft. Taking off from Isla Poí, the aircraft would fly south to Nanawa where they would execute very short landings that were under almost constant attack from the surrounding Bolivian forces. The pilots would park their aircraft as close as possible to the border of the small airstrip where Paraguayan troops would unload the ammunition, and thus the shuttle continued. This series of highly hazardous resupply flights were flown under very gallant circumstances by Tte. Isodoro Jara (s/n 3), Tte.1_ Carmelo Peralta (s/n 4), Cpt. Atilio Ibáñez Rojas (s/n 5) and Cpt. Juan González Doldán (s/n 7) making two additional round-trips each the first day and delivering a total of 1,610 kilos of ammunition. This heroic operation was credited with saving the garrison and preventing it from being overrun by the Bolivians.

The Nanawa operation was not without a price, however, as three of the four Potez that participated sustained severe damage before it was over. Number 3 received a number of small arms hits while it was landing on its third trip but pilot Jara managed to get it to the end of the airstrip so the other aircraft would be clear to land. Number 4 had to make an emergency landing after a hasty take-off from the Nanawa strip between Orihuela and Concepción (possibly due to battle damage) and s/n 5 suffered a similar fate near Isla Poí. Fortunately, the three damaged aircraft were located and, after herculean efforts by ground crews, were sent by ship to Asunción where they enjoyed a much needed complete overhaul. All three returned to active service after several months time, but in the meantime were sorely missed at the front.

During this period, it is also worth noting that no less than the President of Paraguay at the time, Sr. Eusebio Ayala, traveled to the front aboard Potez 25 aircraft when more comfortable transport types were not otherwise available.

Potez 25 T.O.E.'s serials 9 and 15 (the former s/n 8, rebuilt after a major accident) at Isla Poí in 1934 fully marked. Note that, contrary to many previously published accounts, the forward lower engine cowl was left natural metal to aid maintenance crews in keeping the aircraft clean. (Archivo Cnel. Av. (S.R). A. Pasmor).

An excellent close-up view of the twin Madsen machine guns on Paraguayan Potez 25 T.O.E. serial 13 at Isla Poí in 1933. This form of stylized numeral was common on Paraguayan aircraft, whereas Bolivian aircraft tended towards more blocked numerals. (Archivo Cnel. Av. (S.R). A. Pasmor).

With the delivery of Fiat CR.20bis fighters from Italy, the hard-pressed Potez 25s began to enjoy some measure of fighter escort, as the earlier Wibault 73s had proven to be very disappointing (about which more later in this account). On 3 June 1933, six Potez 25s (s/n 9, 10, 11, 12, 13 and 14, the newest aircraft) and two of the Fiats (s/n 11-1 and 11-3) launched from Isla Poí on a mission to bomb the Bolivian fortress of Platanillos, the first really large-scale mission flown by the Aviación en Campaña during the conflict. Each Potez carried 12 bombs of 14 kilos (an overload) to attack the large Bolivian troop concentrations which were threatening the Paraguayan position at Herrera. This mission was a complete success, as the formation virtually destroyed the Bolivian positions at Platanillos with heavy losses suffered by the Bolivian ground forces. For reasons that have not been ascertained, the Paraguayan air attack went completely unmolested by Bolivian air defense, and all of the Paraguayan aircraft regained their base. Unfortunately, Potez 25 s/n 14, piloted by Tte.1° (Naval) Teófilo Fariña Sánchez and Tte.2° Estanislao Arce (gunner), still had two bombs hung up in its wing racks which were found to be oscillating in the slipstream wildly. The crew could not be made aware of this, unfortunately and, upon alighting, the bombs finally were jarred loose and exploded, completely destroying the aircraft. Although both crew members were severely injured, they survived this unfortunate incident.

In August 1933, Potez 25T.O.E.s s/n 12 and 13, escorted by three Fiat CR.20s, conducted an armed reconnaissance-in-force over the Bolivian positions at Falcón, Gondra and Pirizal and, during the remainder of the year, all airworthy Potez 25s conducted intensive reconnaissance and occasional bombing missions over Boliv-

Probably one of the very first Potez 25s to reach Paraguay, this rather casual view shows (probably) serial number 3 at the main Paraguayan aerodrome at Ñu-Guazu in 1932. Note that the aircraft has a conventional Paraguayan roundel under the starboard lower wing (Archivo Cnel. Av. (S.R). A. Pasmor).

ian positions at Herrera, Samaklay, Muñoz and Saavedra. The highly successful Paraguayan offensive at Campo Vía in October 1933 saw much action by the airworthy Potez aircraft of both Escuadrons.

Bolivia took its revenge during the 15 day battle for Cañada Strongest between 10 and 25 May 1934, however, and Potez 25s s/n 7, 10 and 13 were once again used as ammunition transports to resupply the beleaguered Paraguayan garrison, dropping huge ice bars to the 1,500 Paraguayan defenders who were suffering terribly from thirst, certainly a first in aviation history. These very dangerous missions, often flown literally at tree-top level, were flown by Tte. Benito Sánchez Leyton, Luis Tuya (both Uruguayan military aviators who had volunteered to fly combat missions for Paraguay) and Arsenio Vaesken with gunners Tte. Alejandrino Martínez and Job Von Zastrow. In one of these flights, Potez 25 s/n 7 sustained repeated small arms hits but its pilot, Tte. Leyton, managed to reach the airstrip at Cabezón where he landed safely. The ground crew there counted nearly 200 bullet holes in the aircraft and 14 more in Leyton's parachute!

In June 1934, Potez 25s s/n 5, 11 and the re-serialed 15 bombed the Bolivian positions at Cañada El Carmen successfully, where Paraguayan ground forces took nearly 7,000 Bolivian prisoners plus

Crews of the Paraguayan Primera Escuadrilla de Reconocimiento y Bombardeo before a unit Potez 25 T.O.E. at Isla Poí c.1932. You didn't want to stand around in these rigs very long on the ground in the Chaco! (Archivo Cnel. Av. (S.R). A. Pasmor).

Paraguayan crews of the Segunda Escuadrilla de Reconocimiento y Bombardeo standing in full flying gear before a Potez 25 T.O.E. at Isla Poí aerodrome c.1933. The parachutes appear to be predominantly French, but at least one Italian unit is evident (Archivo Cnel. Av. (S.R). A. Pasmor).

large stocks of war materiel. During the course of this battle, perhaps the largest air-to-air action of the war took place when four Paraguayan Potez 25s and two Fiat CR.20s were engaged by a joint Bolivian force of some 11 "Ospreys" and Curtiss "Hawk IIs", lasting nearly 20 minutes. The Potez aircraft, by far the slowest combatants involved, immediately maneuvered to tree-top level during this engagement and, in spite of the intensity of the free-for-all, all Paraguayan aircraft returned to base unscathed.

On 8 July 1934, four Potez 25s (s/n 5, 10, 13 and 15) piloted by Tte. Tuya, Vaesken, Jara and Duarte (with gunners Tte. Martínez, Corbalán, Favio Martínez and Cpt. Von Zastrow) struck the Bolivian fortress at Ballivían from an altitude of 900 meters, dropping 40 bombs between them on the Bolivian airfield near the fortress, where at least seven mixed Curtiss aircraft were observed parked. The Bolivian aircraft were claimed damaged as they apparently could not take off in defense of the garrison. Another 80 bombs were dropped in subsequent missions at other defensive positions around Ballivían with considerable effect, and a fuel depot was destroyed in the process.

One air-to-air combat of note occurred on 12 August 1934, when Potez 25T.O.E. s/n 11, piloted by Cpt. Carmelo Peralta and gunner Tte. Rogelio Etcheverry, while on a patrol over the Florida sector, encountered its arch nemesis, a Bolivian Curtiss-Wright C14R "Osprey" crewed by Maj. Rafael Pabón (pilot) and gunner Sgto. Mario Calvo. Unable to outrun the considerably faster "Osprey," Cpt. Peralta climbed in tight spirals, the crew blasting away with its four Madsen guns whenever the "Osprey" presented a target. Apparently frustrated by this unusual defensive maneuver, Maj. Pabón climbed and dove upon the struggling Potez at least twice. Finally, the "Osprey" made a final dive on the Potez during which it was hit repeatedly by the twin guns of Tte. Etcheverry in the rear cockpit and fell away apparently completely out of control. The "Osprey" indeed plunged into the Chaco and both pilot and gunner were lost. This must certainly rank as one of the final air-to-air combats in which two-place biplane reconnaissance bombers engaged in conflict in this manner in aviation history.

By this time, Paraguayan ground forces were on the advance everywhere in the Chaco and, as a result, other more advanced aerodromes were constructed to place Paraguayan aircraft closer to the rapidly moving front. These included fields at Picuiba, Camacho and Yrendagué and, during September 1934, several of the Potez 25s and two Fiat CR.20s were moved to the advanced field at Picuiba. On 22 September, while flying a reconnaissance mission from there, one of the Potez 25s, escorted by the two Fiats, encountered two Bolivian "Ospreys" enroute (apparently) to Picuiba on a recce of their own. The CR.20s left the Potez to attack the two

One of the unsung heroes of the Paraguayan aviation effort during the Chaco War, (then) Lieutenant Agustin Pasmor in full flying regalia at Concepcion aerodrome enroute to the operating base at Isla Poí in 1932 before a Paraguayan Potez 25. (Archivo Cnel. Av. (S.R). A. Pasmor).

Paraguayan Potez 25 T.O.E. serial number 15 (the rebuilt s/n 8) at Isla Poí c.1934. The aircraft appears to have the rarely seen segmented Paraguayan star/roundel national insignia on the upper wing at this point (Ministerio de Defensa Nacional).

Potez 25 T.O.E. serial 11 returning from a bombing mission. Note that when in flight, the undercarriage of the Potez 25 constricted inwards, another distinctive recognition feature in flight. (Archivo Cnel. Av. (S.R). A. Pasmor)

"Ospreys", which promptly jettisoned their bomb loads and made a high speed escape.

In October 1934, the Paraguayan aviation assets at Picuiba departed there in order to attack Bolivian positions at Lobrego, El Carmen, Carandaiti and near Yrendagué.

By the end of 1934, Bolivian forces, as previously noted, had been all but pushed out of the Chaco Boreal and into the Andean foothills. On 12 December, two Bolivian "Hawk IIs", escorting a Junkers K 43 from the Punta de Alas squadron, were intercepted by an intrepid Potez 25 (s/n 13) flown by Tte. Vaesken and gunner A. Bertoni. The "Hawks" attacked the outclassed Potez vigorously and, after a brief exchange, Tte. Vaesken managed to put his aircraft, now on fire, down in the bush. Miraculously, and to the credit of the Potez, both airmen survived this encounter, but the aircraft was lost.

During the final months of the war in 1935, the remaining air-worthy Potez 25s (now numbering not more than six aircraft) continued offensive missions over Bolivian positions, flying deep in some instances into purely Bolivian territory in the Andean foot-hills. On at least one occasion, they bombed the Bolivian city of Charagua, north of the Rio Parapití, four aircraft participating in this unique attack out of their operating base at Ybamirante, the most advanced Paraguayan operating air base of the war.

The cease-fire finally came on 12 June 1935, but as recently as the day before, four of the surviving force of Paraguayan Potez 25s carried out not fewer than 15 sorties over demoralized Bolivian positions near the front.

Potez 25 T.O.E. serial 13 in-flight during operations near Saavedra in 1933. Again evident are the apparent remnants of the segmented star-roundel national insignia under both lower wing panels (Antonio Sapienza Collection).

Early wartime in-flight formation view of Paraguayan Potez 25 A2s serial 5 and 6 over Ñu-Guazu aerodrome c.1931. Number 6 was shot down by the Bolivian "ace" Rafael Pabon in December 1932. Note the very large presentation of the standard Paraguayan national roundels on the upper wings (Tte. Gonzalo Palau).

Potez 25 A2 serial 5 shown returning from a mission over Nanawa in January 1933. Note the seldom seen underwing "salvo" racks for up to six light bombs each and the standard national insignia roundels (Archivo TCnel Carlos R. Peralta).

Perhaps fittingly, two Potez 25s flew escort to the Paraguayan Breda Ba 44 transport that took the armistice delegation to Villa Montes on 15 June 1935. From first to last, the Potez 25 was the only combatant aircraft type on either side that flew continuously throughout hostilities.

Of the total of 14 Potez 25s acquired by Paraguay, only four airworthy examples survived the war: s/n 5 (a Potez 25 A-2) and Potez 25T.O.E.s s/n 7, 9 and 11. Two were confirmed shot down during the war (s/n 6 and 13), one had been lost in a training accident before formal hostilities commenced (s/n 2) and the remaining seven aircraft were lost through accidents or lack of spare parts with which to return them to service (s/n's 1, 3, 4, 8, 10, 12 and 14). The fate of s/n 15, the rebuilt s/n 8, is unknown.

The four surviving aircraft participated in the Victory Parade organized by the government of President Eusebio Ayala on 20 August 1935, and took a much-deserved place of honor in that fly-past.

After the war, the four survivors still on hand became part of the equipment of the reorganized Primera Escuadrilla de

In-flight view of Potez 25 T.O.E. s/n 12 showing evidence of the field application of the unauthorized "star" national insignia used for a brief period in the zone of operations on some aircraft. It was caught here actually returning from a bombing mission in 1933. Note that the serial number appears under both lower wing panels. (Museo de Historia Militar via A. L. Sapienza).

The rather beefy profile of the Potez 25 T.O.E., compared to the standard Potez 25 A2, is evident in this side view of s/n 13, which appears to have had a name painted on the mid-fuselage (Archivo Cnel Av. (S.R). Agustin Pasmor).

Reconocimiento y Bombardeo based at Campo Grande, which was also known by its Guaraní name, "Ñu-Guazú."

One of the final episodes involving a Paraguayan Potez 25 attested to the faith the type had engendered throughout Paraguayan society. On 7 September 1940, one of the nine year old Potez 25s (s/n 7) was utilized by President (General) José Félix Estigarribia, the hero of the Chaco campaign, to fly himself and his wife (both in the rear gunners cockpit) on his customary 15-minute flight to the popular summer resort at San Bernardino on Lake Ypacaraí, the twin-engined Breda Ba 44 usually employed being inoperable at the time. The pilot, Maj. Carmelo Peralta, the President and his wife, were all killed when the aircraft plunged to earth near the city of Altos. Examination of the remains revealed that the prop shaft had snapped although another contemporary report stated that the aircraft crashed head-long into a hill in very poor visibility.

The three remaining intact Potez 25s (s/n 5, 9 and 11) were placed in reserve until 1943, when they were dismantled. The last airworthy example was s/n 5, one of the original Potez 25 A-2s acquired in 1928!

All of Paraguay's Potez 25s appear to have been painted a very dark green over most of the aircraft (FS34079 seems a close approximation from available evidence) except the lower engine cowling on some aircraft, which apparently was kept free of paint, probably to facilitate the removal of the inevitable oil leaks and stains, as well as corrosive engine coolant leaks. The national colors covered the rudder, with a red star in the white central horizontal stripe. Paraguayan roundels were painted on both outer upper and lower main planes and the individual serial number, in rather stylized white numerals, were painted in large size on the fuselage aft of the gunners position on both sides. After the end of hostilities, all four aircraft remaining were reportedly stripped of paint completely (except national insignia and serials) and the wings and other fabric surfaces doped aluminum. There is some evidence to suggest that an alternative Paraguayan national insignia was carried, at one point, on at least three Potez 25, s/n 4, 5 and 13. This, in the wing positions, consisted of a large red star with a blue circle within its core and white circular central dot. Col. Agustín Pasmor, Maintenance Officer for the Aviación en Campaña during the war, claims these were "unauthorized, merely showing the imaginations of personnel in the field in charge of repainting in isolated instances." Photo evidence also suggests that at least some Potez 25s carried the conventional roundel on both the upper and lower surfaces of the upper wing, as well as the lower surface of the lower wing, as was customary. Some sources also claim that some Paraguayan aircraft had a large yellow star on the fuselage during the war as part of the "national" insignia scheme. This is completely without foundation. It should also be noted that Potez 25 s/n 1 was, late in its career, modified as a two-seat conversion trainer, and served as such with the Escuela de Aviación Militar.

Finally, lest it escape the reader, it should be noted that the Potez 25s were the only combatant Paraguayan recce/bomber air-

Portrait of 1° Tte. Rogelio Etcheverry, the Paraguayan Potez 25 T.O.E. gunner in s/n 11 that shot down Bolivian "ace" Pabon's "Osprey" on 12 August 1934. (Museo de Historia Militar via A. L. Sapienza).

POTEZ
Type 25

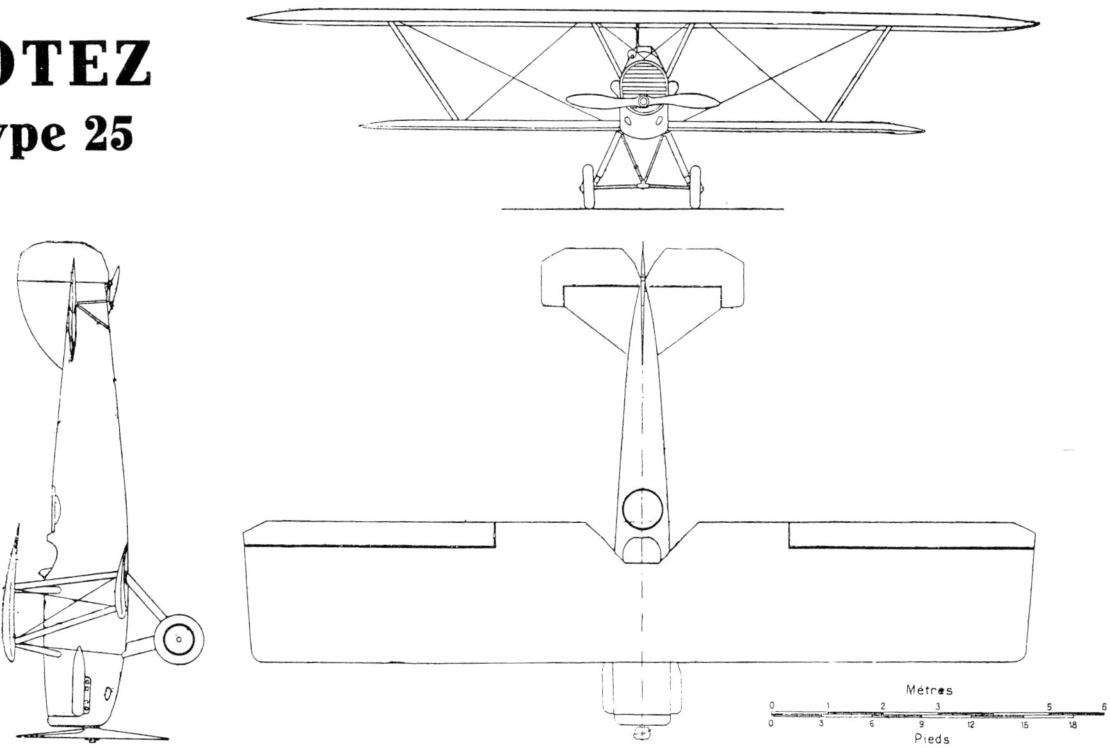

DÉSIGNATION

Potez 25 A2 ou **B. 2**
Disposition de la voilure : biplan.

Moteurs de 350 à 550 HP.
Hélice tractive.

DIMENSIONS

Envergure.... 14m,20 — Longueur.... 9m — Hauteur.... 3m,30

Surfaces

Voilure	Plan supérieur	34m²
	Plan inférieur	12m²
Voilure totale		46m²
Ailerons		4m²

Empennage horizontal : plan fixe	2m²,20
Gouvernail de profondeur	2m²,20
Gouvernail de direction	1m²,05
Dérive	1m²,15

Répartition des poids
(cas du Lorraine 450 HP)

		Indices
Planeur	615kg	0,31
Groupe-moteur (partie amovible)	530kg	0,27
Poids à vide	1145kg	0,58
Charge normale	835kg	0,42
Poids total normal	1980kg	1,00
Charge utile maxima	1255kg	0,52
Poids maximum en ordre de vol	2400kg	1.00

Poids par HP	normal	4kg,20
	maximum	5kg,70
Poids par mètre carré	normal	43kg
	maximum	52kg
HP par mètre carré		10,20

Particularités

Le groupe moteur complet, avec radiateur et réservoir d'huile, est monté sur une poutre indépendante qui s'attache à l'avant du fuselage d'une façon simple.

L'adaptation est faite pour les moteurs suivants : *Lorraine* 450 HP, *Jupiter* 420 HP, *Salmson* 500 HP, *Renault* 480 HP, *Farman* 500 HP, *Hispano* 450/500 HP en V et W, *Jaguar* 350 HP, etc.

Double commande. Réglage du plan fixe en vol. Châssis d'atterrissage à jambes élastiques, système "**Potez**", supprimant les sandows.

Construction

Mixte. Ailes en bois, entoilées. Fuselage carrossé en contreplaqué. Bâti-moteur et train d'atterrissage métalliques.

Destination

Biplace militaire de reconnaissance ou de bombardement de jour.

Three-view drawing of the standard Potez 25 A2, the configuration of the first six aircraft acquired by Paraguay ("L'Aeronautique").

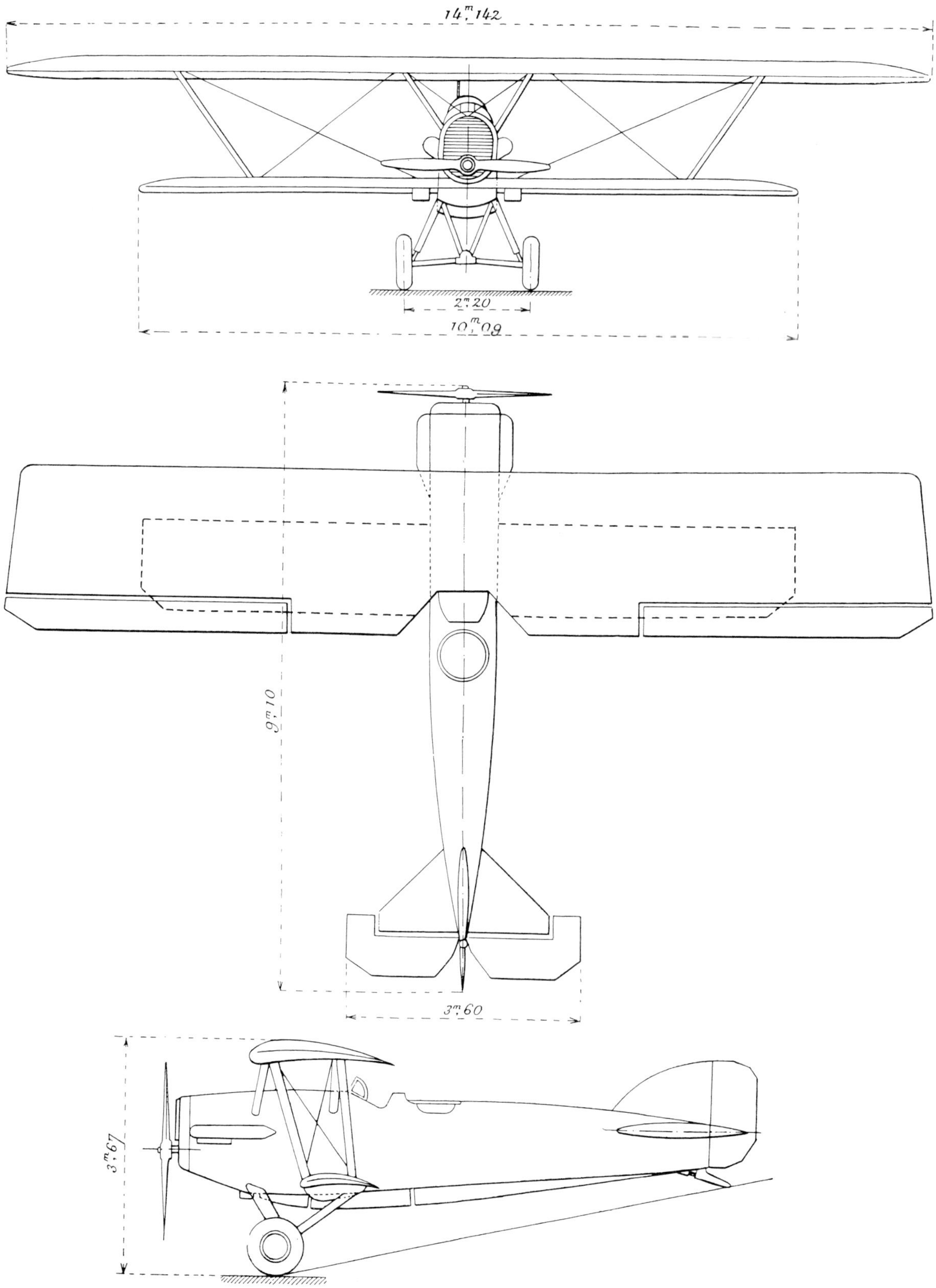
14m142
2m20
10m09
9m10
3m60
3m67

Note the presentation of the type designation and individual aircraft serial on the rudder of the Potez 25 A2 on the right while the very first Paraguayan Potez 25, s/n 1, serves as backdrop for a group of military personnel in a hangar at Ñu-Guazu in 1929. (Museo de Historia Militar via A. L. Sapienza).

craft actively employed as such during the entire Chaco campaign, while Bolivia enjoyed the versatile services of not fewer than six major types - the Curtiss-Wright C14R "Osprey," the Breguet 19 A-2s, Vickers Type 149 "Vespa IIIs", Junkers K 43hs, Fokker C.Vbs, and Curtiss Cyclone "Falcons"!

This account cannot end without presenting the roster of the airmen of the Escuadrilla de Reconocimiento y Bombardeo which crewed these vintage aircraft against such odds:

Maj. José Atilio Migone
Cap. Leandro Aponte
Cap. Trifón Benítez Vera
Cap. José M. Fernández
Cap. Víctor A. Vallejo R.
Cap. Ramón Della Loggia
Cap. Juan González Doldán
Cap. Atilio Ibañéz Rojas
Tte.1° (Marina) Víctor Urbieta Rojas
Cap. Román Garcia
Cap. Walter Gwynn
Cap. Isodoro Jara
Cap. Tomás A. Ruffinelli
Cap. Cármelo Peralta
Cap. Emilio Rocholl
Tte.1° Abel Vera y Aragón
Tte.1° César Abente Benítez
Tte.1° (HC) Luis Tuya
Tte.1° (HC) Benito Sánchez Leiton
Tte.1° Arsenio Vaesken
Tte.2° (Marina) Teófilo Fariña Sanchez
Guardiamarina (PAM) Enrique Planás
Tte.2° Orlando Salerno Netto
Tte.2° Homero Duarte
Tte.2° Abelardo Bertoni
Tte.2° Abdón Caballero Alvarez
Tte.2° Fernando Pérez Veneri
Tte.2° Hermes Gómez Lezcano
Tte.2° Juan P. Pedretti
Tte.2° Luis Ugarriza
Tte.2° Gonzalo Samaniego Abente

Finally, this account of the use of Potez 25s by Paraguay cannot be complete without mention of the fact that Bolivia, also, apparently acquired at least one Potez reconnaissance bomber as late as 1926. This aircraft, reliably reported in several French periodicals of the time, as well as JANE'S, was possibly a Potez XV, but could also have been an early Potez XXV. Apparently acquired for evaluation purposes at the urging of the French Mission, the fate of the aircraft is unknown.

OPPOSITE: Three-view of the Potez 25 T.O.E., which, with its much greater fuel capacity, extended the reach of the otherwise thinly spread Paraguayan aviation assets during the war (Potez).

VICKERS TYPE 149 "VESPA III"

Certainly a candidate for the list of all time ugliest aircraft, the ungainly looking Vickers Type 149 "Vespa III" actually rendered valuable service to Bolivia during the Chaco War, out of all proportion to the numbers involved.

Part of a comprehensive arms package ordered from the Vickers arms conglomerate, originally, in 1927 as part of the $27,000,000 Dillon Read "development" loan to the Bolivian Government, a total of six "Vespa IIIs" is normally quoted as having been delivered to Bolivia between 23 March (when the first one was erected by C. B. Holmes at La Paz) and 25 July 1929 on Vickers Order No.65-A at a cost of 34,340 Pounds Sterling. The reliable mating of the Bristol "Jupiter" VI engine of 450hp with lighter and stiffer metal wings produced an aircraft with excellent high-lift properties and, especially important to Bolivia, a high operational ceiling. These features, coupled with short take-off and landing characteristics, apparently impressed a Bolivian purchasing mission which visited England in 1927.

Indeed, the high-altitude capabilities of the "Vespa" mark it as the first military aircraft acquired by Bolivia from the outset tailored for operations from its primarily lofty airfields, particularly "El Alto" at La Paz. Shortly after delivery, RAF Flying Officer Banting, who apparently accompanied the first three "Vespas" on delivery, flew with a passenger from "El Alto" to the region of the mountain Illimani (24,200 feet) and to Illampu Lake (25,000 feet) with an unsupercharged engine and without the benefit of oxygen. Banting also had responsibility for transitioning Bolivian pilots to the new aircraft, which was not an easy task as the aircraft did not have dual controls. Most training was conducted in the "Vendace" trainers, then Banting would take the Bolivian pilots up one at a time while he undoubtedly suffered in silence in the back seat!

The "Vespa" also had other very attractive characteristics. It was light and sensitive to the controls and, for an aircraft of its size, had good maneuverability, and even at high altitudes could maintain control right down to stalling speeds. It also could carry a variety of ordnance and operational loads, and had ample fuselage space for various configurations. It was also considered relatively easy to maintain and replacement parts could be installed with satisfactory rapidity. Last but not least, the aircraft was considered very "crashworthy" and offered unusual safety for the crew in the event of a crash.

The one recurring problem with the Bolivian "Vespas," which, aside from an area just aft of the engine which was painted matt black, were natural metal or doped aluminum overall, was to find a suitable protective finish to cover the Duralumin - either in conjunction with or, preferably, in place of anodic treatment, that would stand up to the wear and tear of service in the harsh Bolivian extremes. The paints carried on the aircraft at delivery (Nitro-Cellulose) tended to chip off the Duralumin surfaces very quickly and was also insufficiently resistent to the attack of liquid petrol fuels, which were inevitably splashed and splattered during field refuel-

Certainly a candidate for ugliest aircraft of all time, the Vickers Type 149 "Vespa III" actually served Bolivia very well. More durable than its appearance would suggest, the type served completely through the Chaco War period in dwindling numbers. In this view, the Vickers pilot, W. R. Banting, is shown with Sr. Siemel (with camera) prior to take-off on a 25,000 foot flight over Lake Illampu (courtesy E. B. Morgan)

Good detail view of a Bolivian "Vespa III," serial number 5 at El Alto aerodrome in 1930. One shortcoming of the design was the rather unfortunate view of the pilot, who must have felt most claustrophobic in his cockpit! (via Ramiro Molina Alanes).

ing. No truly satisfactory solution was found for the Bolivian "Vespas" but, as it turned out, they were attrited before the problem could become chronic.

Although the total of complete "Vespa IIIs" delivered to Bolivia has always been correctly stated as six aircraft, it is not generally known that the original contract also included provision of "three 'spare' airframes." Apparently these "spare" airframes were, in fact, erected and employed by Bolivia, as a total of nine individual serial numbers are known (s/n "1" to "6" for the original "complete" aircraft, sometimes given as Roman numerals, which may be correct for the early examples, and "18," "23" and "44" for the "spares"). It has also been suggested that these last three serials were "new" serials issued to rebuilt aircraft from the initial batch of six (usually cited as "1", "2" and "4"), but a U.S. Military Intelligence Report from the U.S. Legation in La Paz dated 23 May 1933 cited Bolivia as having acquired "nine Vickers Vespas in 1929." This seems to be corroborated by a report filed on 20 January 1932 by 1LT A. Y. Smith, U.S. Army Air Corps, who was traveling in South America at that time and who rendered a detailed report on his observations to the War Department upon his return to the U.S. 1LT Smith reported that Bolivia was operating "...seven Vickers observation-bombers at El Alto at the time..."

Bolivia's Cuerpo de Aviación operated its "Vespas" in conjunction with surviving examples of its Fokker C-Vbs and Breguet 19 A-2s, initially, as part of the Grupo de Combate. As Bolivia had the skies over the Chaco region virtually to itself from 1928 into 1932, the trio of types were hazarded only by occasional ground fire. The initial Bolivian thrust in June 1932 took the Paraguayan defenders by surprise, and, by this time, the "Vespas" had been organized into a unit designated as the Escuadrilla de Reconocimiento Aereo. This was commanded by Cpt. Luis Paravicini and included pilots Tte. Juan Antonio Rivera and Elias Belmonte, and gunner/mechanics Sub.Tte. Alejandro Robles, Sub.Ofc. José Ardiles and Lucio Bustillos, which would seem to indicate that three of the six "Vespas" remained operational by this time. The unit deployed to Villa Montes on 18 July 1932 to commence operations in the Chaco area.

Although the "Vespas" and accompanying Vickers "Bolivian Scout" fighters of the Grupo de Combate detachment at Villa Montes carried out a number of missions in support of infantry forces, these were at first poorly coordinated and, in the words of an official Bolivian history, "taught many lessons in air-ground cooperation that would prove crucial in the three years of combat that lay ahead."

"Vespas" were active in the battles for Boquerón and Toledo (including s/n "6") and, in the course of one of these missions, a "Vespa" flown by Tte. Elias Belmonte and his gunner, Julio A. Saavedra Goytia, suffered anti-aircraft hits while leaving their target and, enroute back to their operating base at Muñoz, made a forced landing near Fortín Aquino, and both crew were made prisoners of war. This appears to have been the first aircraft, of either side, lost during hostilities and was apparently s/n "5" which was reported lost on either 27 or 31 July 1932. This may have been the "Vespa" recovered by Paraguayan forces and, with interim field repairs, was interned in Argentina when the Paraguayan pilot, attempting to fly it home to Asunción, became lost. The fate of the aircraft is unknown, although it is believed that Argentina returned it to Bolivian control.

Bolivian operations with "Vespas" continued, however, and usually included long-range, solitary reconnaissance missions over Paraguayan positions, including at least one to the Isla Poí Paraguayan Headquarters region, where a "Vespa" flown by Cpt. Luis Ernst Rivera attacked columns of Paraguayan ground forces on the march on 7 September 1932 solo.

"Vespas," now apparently reduced to but two airworthy examples, also carried out valuable work during the retrograde operation known to the Bolivians as the battle of "Kilometro 7 de Saavedra" between 25 October and 20 December, providing some limited aerial protection to the retreating ground troops in the Boquerón-Arce-Yujra-Alihuatá-Saavedra sector. Indeed, the intelligence provided to the Bolivian leadership by recon flights made by Maj. Jordán and observer Col. Toro were credited with providing the essential information needed for the Bolivian defense at Saavedra.

At this juncture, apparently all remaining "Vespas" and "Scouts" were moved from La Paz to the front, as Maj. Jordán was able to report that he had "...12 Vickers aircraft at Muñoz" for duty at the front.

By January 1933, the first of the Curtiss "Hawk II" fighters and "Osprey" two-place combat aircraft had arrived at the front, enabling the crews of the now rather tired Vickers aircraft an opportunity to attend to their stressed aircraft. The older, veteran pilots quickly transitioned to the new Curtiss mounts and, for the moment, operations with the "Vespas" came to a virtual halt, as there were few crews to man them.

A Bolivian "Vespa III" at altitude over the mountainous interior of the country, for which it was best suited. Operations in the Chaco were at very low altitude, comparatively, but the immense wing area of the aircraft made it a very stable bombing platform (courtesy E. B. Morgan).

With the graduation of 12 new pilots at the Escuela de Aviación, however, crews became available to once again operate the Vickers. However, their inexperience shortly manifested itself when one of the new pilots and his observer, Sub.Tte. René Dorado and Maj. Eliodoro Nery, were lost in the crash of one of the "Vespas" during a familiarization flight in the vicinity of San Antonio at the time of the defense of Ballivian circa 8 July 1934.

From this point on, the remaining "Vespas" were apparently seconded to other duties, and at least one was returned to La Paz for overhaul as, upon its return to the front to aid in the defense of Carandaiti, it was crashed on landing and was a complete write-off.

Other known losses of Bolivian "Vespas" include one which apparently was a write-off at Santa Cruz some time in 1930 before open hostilities commenced and one claimed shot down by Paraguayan anti-aircraft fire on 25 February 1933, although official Bolivian loss figures do not admit to this casualty. At least two "Vespas" survived to be withdrawn from use near the end of 1935, and, although JANE'S All the World's Aircraft for 1940 still reported the type on strength with Bolivia's air arm, this is not supported by existing intelligence data.

For the record, Bolivia officially cites the following pilots of the Escuadrilla de Reconocimiento as having flown in operations during the Chaco War on "Vespa" aircraft:

Cap. Luis Paravicini
Cap. Juan Antonio Rivera
Tte. Elias Belmonte

Bolivian crews in the area of operations before a pair of "Vespa IIIs" and displaying two of the largest caliber bombs used during the war, one of which could be carried at a time by a "Vespa" under the mid-fuselage (via Georg von Roitberg).

Color artists concept of one of the Bolivian Junkers K 43 bombers. This configuration is, in fact, the way that at least one of the aircraft appeared after the end of the war, as none of the aircraft are known to have had engine "speed rings" during the conflict. (Courtesy Gral. Div. Ae. (Sp) Alberto Paz Soldan Pol).

JUNKERS K 43h

Due to the secrecy surrounding the marketing of Junkers aircraft by the Swedish AB Flygindustri of Limhamn, Sweden between the wars, the Junkers K 43h light bombardment and reconnaissance aircraft acquired by Bolivia are cloaked in mysterious circumstances.

Indeed, it is not easy to find information of any kind regarding these little-known aircraft, developments of the W 34 transport. First revealed in 1927, the aircraft employed the distinctive Junkers corrugated skinning and, initially, had a virtually identical structure to that of its commercial counterpart. The K 43 had empty and loaded weights of 2,425 pounds (1,100kg) and 3,527 pounds (1,600kg) respectively and average performance, depending on engine selected, ranged from a high speed of 150mph (242 km/h) to cruise of 132mph (212 km/h). Armament consisted of one fixed, forward firing synchronized 7.62mm machine gun with a similar weapon or pair of weapons in an open dorsal position, and up to six 110 pound (50kg) bombs slung from external racks.

Production K 43s could take engines from 480hp to 750hp (the Bolivian aircraft apparently had 660hp BMW engines) and, in the case of the aircraft Bolivia acquired, the normal crew could be increased to three, a second open dorsal gun position being available, although apparently seldom actually used, if at all, due to the sacrifice in bomb load. Range, with a crew of two, two guns and maximum bomb load, which was of interest to the Bolivians, was 528 miles (850km). K 43s were acquired by the Finnish, Argentine, and Colombian air arms and the Portuguese Navy, besides those acquired by Bolivia.

The first known mention of Junkers combat aircraft for Bolivia surfaced in late September and October 1933 when, in the course of correspondence home to corporate headquarters, the Curtiss-Wright Export Corporation representatives in La Paz, Owen Shannon and "Cliff" Travis, reported that, in order to "satisfy General Kundt and the Lloyd Aereo Boliviano, orders....may have to be split up and possibly three out of the 10 ships may have to go to

Actual wartime view of Bolivian Junkers K 43 s/n 103 (with 102 in the background) showing the obvious light gray undersurfaces and dark green upper coloring. The four underwing bomb racks are clearly visible (the aircraft had no internal racks). Left to right are Cap. Eliodoro Nery, Maj. Jorge Jordan and Tte. J. Antonio Rivera, during the struggle for Fortin Ballivian in 1934. This may be the elusive, third K43 not previously confirmed. (via Ramiro Molina Alanes).

Junkers, although our performance [in discussing Curtiss Cyclone Falcons] is better." An October letter went on to report that "...three Junkers bombers have definitely been purchased, but we had no chance at that business, as LAB is in the deal and have agreed to take the planes for their line after the war. They are convertible jobs and Junkers is standard equipment on the LAB." By 15 December, the exact types in question had apparently swung between three single-engined K 43s or, alternatively, two Junkers Ju 52/3ms outfitted as bombers, as Bolivia was bent on obtaining really long-range heavy bombers at the time that could reach Asunción (see the section on the Curtiss-Wright "Condors" in the chapter Things that Didn't Happen).

A German diplomatic report dated 24 March 1934 stated that "...the Bolivian Army has acquired three Junkers K 43s" although the report did not specify whether the aircraft had arrived yet or not. They apparently had not, since if they had, they certainly would have been pressed into service as quickly as possible, in view of the situation at the front.

The reader should be cautioned not to confuse the K 43s with the very similar W 34cis used during the war, primarily as transports but, on at least several occasions, as bombers (see Transport chapter).

As nearly as can be determined, only two K 43hs have actually been traced to Bolivia (although even the officially sanctioned Bolivian Air Force history of the conflict, Conduccion de la Fuerza Aerea Boliviana en la Guerra del Chaco states that three were acquired by January 1935) and they seem to have reached the front by September 1934, when at least one was crewed by T. W. Smith, author of Gran Chaco Adventure (required reading for anyone interested in gaining a sense of what life in the Chaco was like during the war) in a bombing attack on Carandaiti late in that month, although this was somewhat inconclusive. These aircraft were most likely c/n 2755 and 2756, and received Bolivian serials "104" and "105." The manner of delivery is a matter of speculation, but it appears that they were flight-delivered to Bolivia from Brazil, after being shipped there and erected. It has also been speculated that the

Officers of the High Command at Villa Montes pose before one of the highly prized Junkers K 43s in 1934. In this view, the aircraft bears clear evidence of a mottled camouflage scheme on the upper surfaces, rare during the conflict. (via Ramiro Molina Alanes).

German-run SCADTA airline of Colombia may have had some involvement in their delivery and assembly. These aircraft were amongst the very few Bolivian aircraft that bore camouflage from the start, and they appear to have arrived with dark green (some sources say "rusty green") upper surfaces with either a light gray or light blue underside. Standard Bolivian roundels were carried in all six positions, and the three digit serials were worn on the vertical fin in about 18 inch black numerals.

The K 43s were apparently assigned immediately to what came to be known as the Punta de Alas ("Pointed Wing") squadron, alongside a mix of Junkers W 34 aircraft that had been converted to mount bomb racks, as well as at least one C14R "Osprey." So far as can be determined, the first action in which the K 43s were employed in earnest was during the combat for El Carmen between 13 and 22 November 1934. On 14 November, these were reported to have made their baptism of fire in the company of an "Osprey" (which spotted targets for them) when they flew a bombing mission from Currurenda to Comachu. On the short-range missions flown by the Punta de Alas squadron, the K 43hs usually carried only a crew of two, minimal fuel load, and a bomb load of four 200 pound bombs. In one action sometime around Christmas 1934, one of the K 43s was reportedly attacked by a Paraguayan Potez 25 (a most intrepid interception, if true), and the K 43 was reported by one of the crew members, none other than Thomas W. Smith, the Englishman, to have possibly damaged the Paraguayan aircraft with its forward firing gun in the melee!

A possible clue to the fate of the alleged third K 43h (probably s/n 103) may rest in the reported loss of a single-engined Bolivian Junkers, with its three-man crew, in September 1934 on the Picuiba front, about seven miles from Carandaiti. As all other W 34s seem to be accounted for, this may in fact have been the "missing" third K 43h.

With the arrival of the armistice, two K 43hs definitely were still on hand and these apparently were eventually passed into the hands of Lloyd Aeréo Boliviano where they were stripped of armament and bomb racks and converted into conventional airliners,

Rear gunner posing with his .30 caliber weapon aboard one of the Bolivian Junkers K 43 bombers c.1935. Although the aircraft could also carry a fixed forward-firing gun and a ventral tunnel gun, there is no evidence the latter was ever employed. (via Ramiro Molina Alanes).

bearing the LAB names Sajama and Mururata (sometimes given as Murutata). Sajama was written off in an accident at Cuybaja with LAB on 10 or 13 March 1937 (at which time it was given as a W 33) but Mururata remained in service with LAB as late as October 1938 and probably beyond.

Bolivian pilots known to have been involved in W 34 and K 43 operations included:

Cap. Luis Paravicini
Cap. Juan Antonio Rivera
Tte. Fabían Monasterios
Tte. Alberto Alarcón
Tte. Emilio Beltrán

CURTISS "CYCLONE FALCON"

Bolivia's first introduction to a Curtiss "Falcon" came in February 1928, when none other than Jimmy Doolittle, in company with Curtiss pilot Bill McMullen, demonstrated a Curtiss P-1A "Hawk" and a Curtiss D-12 (in-line engined) "Falcon" at the Escuela de Aviación at "El Alto." During the demonstration, which pitted one of the relatively new Bolivian Breguet 19s (named "Potosi") piloted by the French Mission chief, Major Lemaitre in "mock" combat for 20 minutes with the "Hawk" and "Falcon," the Swiss advisor, Major Haeberli, judged the contest a "draw" but the exhibition none the less made a lasting impression on the Bolivian leadership. This event has undoubtedly led to the frequent reports that Bolivia acquired D-12 powered examples of both the "Hawk" and "Falcon," which most assuredly did not occur. However, Curtiss did make an effort at this early juncture to interest the Bolivians in the two types, when Boyd F. Sherman of Curtiss arrived in La Paz in late October 1927 to attempt to interest the Government in a mix of 20 of the aircraft (apparently 10 of each). However, in the words of the U.S. Legation Chief at the time, "when he learned of the enormity of the Bolivians order with Vickers for arms and aircraft of all kinds, he left in disgust." It is interesting to note that the Curtiss party that demonstrated the "Hawk" and "Falcon" in La Paz went on to demonstrate the same aircraft Chile, Argentina, Brazil - and Paraguay!

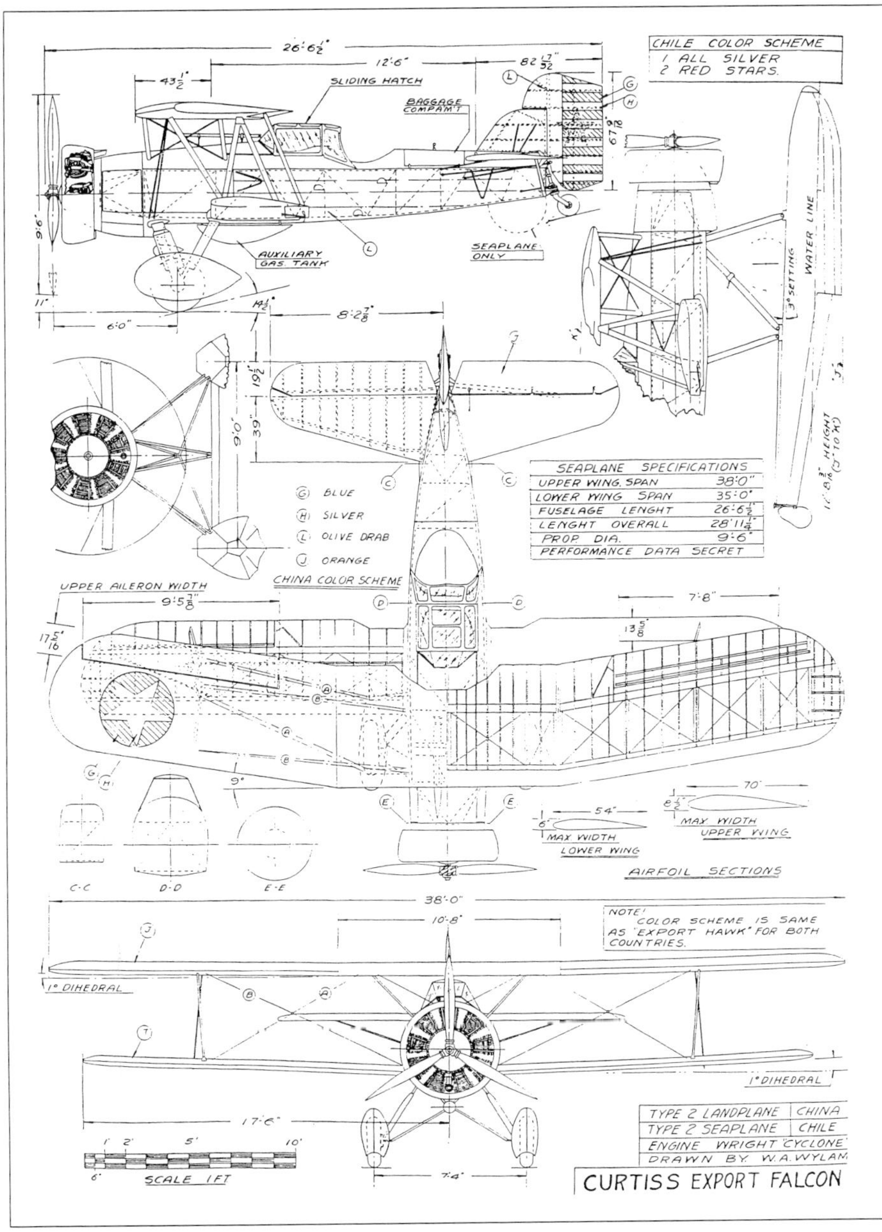

A surprisingly accurate three-view of one of the truly classic aircraft to reach Bolivia during the Chaco War, the Curtiss "Cyclone Falcon." Operated in the field both with and without the wheel spats and complete canopy, the aircraft arrived late and saw comparatively little action (courtesy W. A. Wylam).

A "Falcon" connection with Bolivia next occurred in early February 1933, when Curtiss was attempting to close out its costly "factory" set-up at Los Cerrillos in neighboring Santiago, Chile. While some numbers of P-1A and P-1B "Hawk" variants and D-12 powered "Falcons" had been sold to Chile as a result of this arrangement, several were left over, and one D-12 "Falcon" and one "Hawk" were offered to Bolivia "for a certain price." Apparently the Bolivians didn't like "the price," and the aircraft instead ended up in Peru.

The first documented introduction of the Curtiss "Cyclone Falcon" into the Bolivian scenario came in a communication from C. W. Webster, President of Curtiss-Wright Export Corp., to CPT "Cliff" Travis, Curtiss' representative in La Paz dated 30 May 1933. In that letter, Webster sent Travis photos of the newly developed aircraft and detailed information on its performance (particularly at high altitudes), and a gentle prod to "sell" the aircraft to the Bolivians. Travis replied that "he had already started talking 'Falcon'" but, as noted earlier in the description of the C14R "Osprey," he felt that "the price will probably not appeal to them," and the Bolivians felt they were getting all they needed at the moment from the durable (and cheaper) "Ospreys."

It is of interest to note that at this time Travis was authorized to quote on the "Falcons" with ar-

On this Bolivian "Cyclone Falcon," the national colors have been extended completely forward on the upper rudder and no individual serial has been assigned. The canopy, in an open position, slid forward as shown, giving a rather unusual appearance. One of the officers pictured is Tte. Claudio Lopez Cordova, and the picture was taken at El Alto in 1936, after the end of the war (via Ramiro Molina Alanes).

mament ($28,500 f.o.b.) and without armament ($26,400 f.o.b.). Travis felt that it was vital to demonstrate one of the aircraft to the Bolivians first hand. However, this presented Curtiss-Wright Export with a rather delicate problem: the only "Cyclone Falcon" remotely close to Bolivia was, as of 25 July 1933, in Buenos Aires, Argentina, being demonstrated to Argentine Army and Naval aviation authorities. As Curtiss was distinctly aware that Argentina favored Paraguay in the on-going conflict, Webster cautioned Travis to "feel out Colonel Zuloaga on the idea of getting permission to demonstrate this job in La Paz...but do nothing whatsoever to alter our good relations with the Argentine Government, and bear in mind that we wish to sell this ship to Argentina along with a contract for the purchase of manufacturing rights." Travis was no doubt earning his money!

As the well-liked "Ospreys" were attrited in the Chaco, it became clear to the Bolivians that, indeed, a heavier and more capable aircraft would have to be ordered to replace the lost aircraft, and Curtiss-Wright Export Corp. finally secured an order for nine Curtiss "Cyclone Falcons" on Sales Order No.132. These were Curtiss constructors numbers 11815 to 11823 (which also had "type" Constructors Numbers F-SF-17 to F-SF-25). The last of these was weighed for delivery at Buffalo on 10 July 1934 and the first had arrived at "El Alto" via the Chilean port of Arica by September 1934.

This wartime view of an un-serialed "Cyclone Falcon" at Villa Montes in 1934 shows the national colors on the rudder in a standard vertical presentation. The "Falcon" was one of the first Bolivian aircraft to have the luxury of a tail wheel (via Ramiro Molina Alanes).

Evidence that Bolivia's "Cyclone Falcons" used the rarely seen ventral auxiliary fuel tank and wheel spats. This example, photographed at Villa Montes in 1935, is also noteworthy in that the national color apparently only occupied the upper-half of the rudder. Compare this image to the next photo (via Antonio L. Sapienza).

The "Bolivian Falcons" (as they were referred to in Curtiss documents) were somewhat different than the Colombian "Cyclone Falcons," to which they bore a strong resemblance. The Bolivian aircraft had only one forward firing .30 caliber machine gun and a similar gun for the rear gunner (the Colombian aircraft also had two wing mounted guns).

The Bolivian "Cyclone Falcons" were assigned serials 201 to 209 (although one of these was apparently not painted on the aircraft). Additionally, a number of the aircraft had slight equipment differences. For example, at least two of the aircraft had the peculiar half-cockpit canopy over the pilots cockpit which, when open, slid forward rather than aft, making a most unusual appearance. At least two others did not have the canopy at all, but standard windscreens on the forward edge of both front and rear cockpits. In these aircraft, there was no evidence of a ring-mount for a rear gun.

Attrition on the "Falcons" was rather severe, as these aircraft proved something of a handful for the Bolivian pilots who (for the most part) transitioned into them from old Fokker C-Vbs, Breguet

This photo proves that Bolivia received at least two distinctly different variants of the "Cyclone Falcon." Note the differences between the undercarriage members, belly slip-tank, and standard wind-screen on this example at Villa Montes during the war. (via Ramiro Molina Alanes).

Paraguay also operated a solitary Curtiss "Falcon," identical to this sister ship shown in Brazilian colors. One of the Curtiss Chilean "factory" assembled aircraft flight delivered across the Chaco war zone in 1932, the Paraguayans "interned" one when it made a forced landing and, usually minus armament, used it as a high-speed transport (Hagedorn Collection).

19s and Curtiss-Wright C14R "Ospreys." Colonel Jordan himself, by then the Commander of the Cuerpo de Aviación, crashed one of the newly delivered machines at "El Alto" in September 1934 on a "test flight" during which he performed aerobatics over the high-altitude field at "extremely low altitude." He walked away from the wreckage, reportedly "much impressed with the performance of the 'Falcon'"!

As crews were transitioned to the new aircraft and as they were slowly assembled, they were dispatched piecemeal to Villa Montes to join the several "escuadrillas" that had been formed with mixed equipment. As nearly as can be determined from existing records, at no time were more than five of the "Falcons" in the Chaco airworthy at any one time.

Another "Falcon" was lost 12 December 1934 when Tte. Ricardo Rocha and his observer, Sub.Tte. Heraclio Melgar, were killed in the unexplained crash of their aircraft near Villa Montes. It was suspected that there had been a prop failure (the "Falcons" were equipped with three-bladed props, which were a novelty to the Bolivians at the time). By Christmas, another "Falcon" was unserviceable at Villa Montes with prop problems, seeming to confirm the cause of the earlier loss.

One of the largest Bolivian "raids" of the war took place on 24 January 1935, when 11 aircraft (reportedly consisting of all three Junkers K 43hs, two "Ospreys," three "Hawks" a venerable Vickers "Scout" and two "Falcons") struck at Paraguayan positions in force. A more unlikely formation would be hard to imagine!

Two further "Falcons" were lost in the intensive operations in defense of the shrinking Bolivian defense perimeter going into April 1935. In one of these, Tte. Luis Paredes was lost (he was apparently flying solo) and, in the other, Tte. Alberto Montaño and his observer Tte. Nicolás Rojas, died at Villa Montes as they attempted

Never before illustrated, Curtiss D-12 "Falcon" s/n 17 of the Paraguayan air arm, showing evidence of its dual 7.7mm guns mounted in the rear cockpit. It was photographed at the Isla Poí aerodrome sometime in 1933. (Museo de Historia Militar via A. L. Sapienza).

to land delivering the last of the "Falcons" to the front. Thus, of the nine "Falcons" obtained in late 1934, at least four were lost to accidents alone.

A word about the "Cyclone Falcon" performance is in order at this point, as the type remains so obscure to this day that few details about this aesthetically pleasing attack aircraft have ever been published. Delivered with the classic wheel pants, like their Colombian brethren, the Bolivian "Falcons" had the 712hp Wright SR-1820F-2 "Cyclone" engine (hence its name, to distinguish it from the essentially similar D-12 powered "Falcon"). The aircraft could carry up to 440 pounds of bombs (normally consisting of four 110 pounders) on wing racks and could make 178mph top speed - not too bad for so heavy an aircraft (5,734 pounds gross weight) and had a service ceiling of 23,900 feet (7,285m). In the Chaco, where the aircraft were often flown very light and with only the pilot aboard, the performance was said to be considerably better. The stylish wheel "pants" were apparently dispensed with in a number of cases.

At least two of Bolivia's "Falcons" survived the Chaco war and, incredibly, survived on the Bolivian Air Order-of-Battle (AOB) as late as 1 January 1949, when one was still listed (although it was non-operable at the time). Two were in use throughout the WW2 years intermittently, apparently as part-time target tugs. As of 10 April 1946, both were still on hand at "El Alto" with the 1° Escuadron de Entrenamiento, listed a "third-line (obsolete) equipment." As in the case of the surviving Osprey's, Hawks and CW-16E's, Bolivia requested Lend-Lease funding to effect repairs on two Falcons on 19 March 1945, but this was apparently not approved.

Bolivian pilots official credited with serving on "Falcons" included:

Cap. Juan Antonio Rivera
Tte. Alberto Paz Soldán
Tte. Emilio Beltrán
Tte. Fabián Monasterios
Tte. José Chacón
Tte. Alberto Montaño
Tte. Luis Paredes

Another view of the solitary Paraguayan Curtiss D-12 "Falcon" s/n 17 with a host of Paraguayan personnel at the Ñu-Guazu aerodrome in 1933. It is not clear whether the aircraft was painted white or doped aluminum overall. (Museo de Historia Militar via A. L. Sapienza).

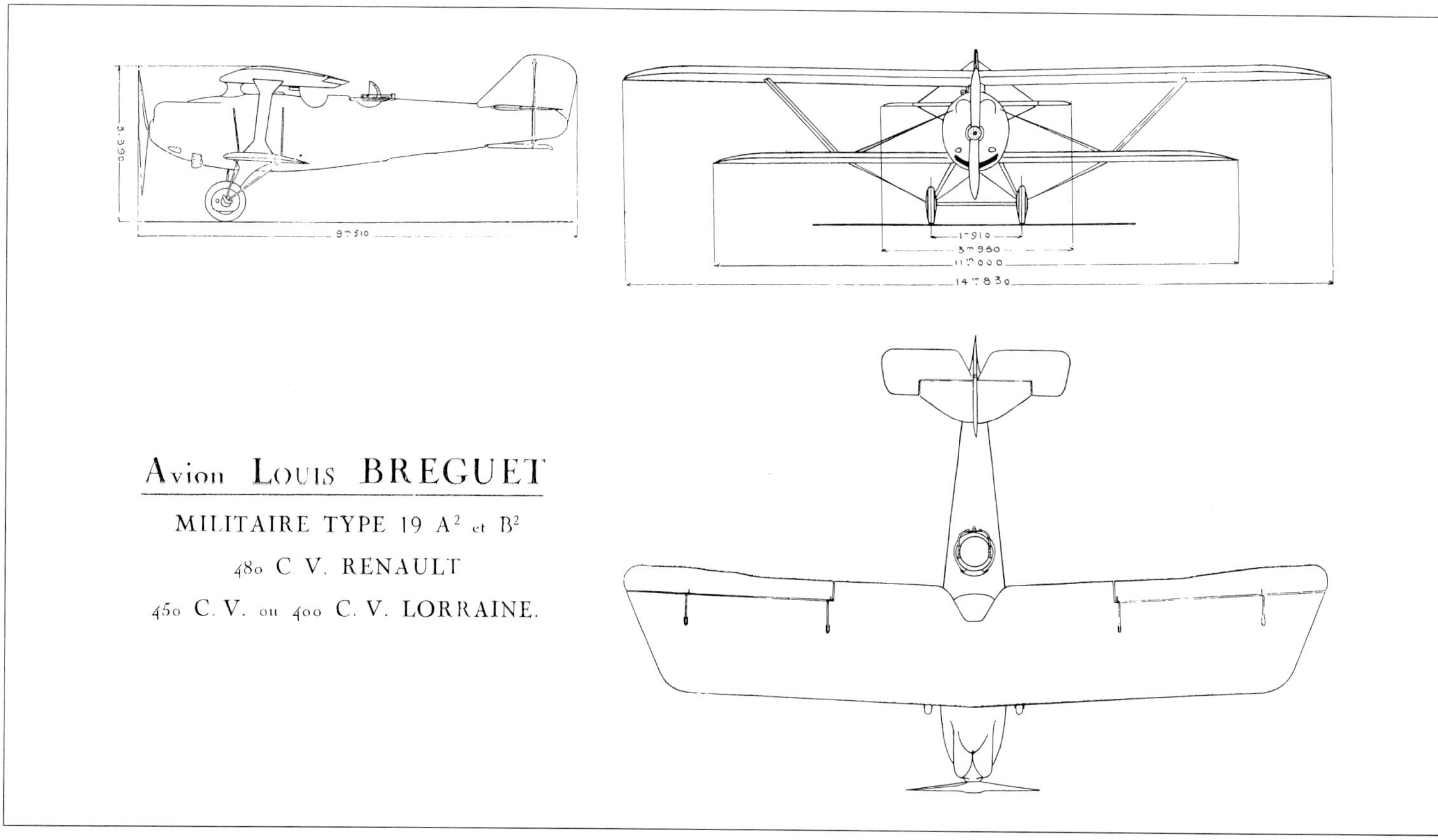

Three-view drawing of the Breguet 19 A2, one of the trio of recon-bomber types that Bolivia had standardized upon at the start of the conflict (Breguet).

BREGUET 19

The Breguet XIX series of light bombardment and Army co-operation aircraft need little introduction. First introduced in 1921 as a successor to the successful World War One era Breguet XIV, more than a thousand Breguet 19s (as they had become known in French literature by February 1928) were built and exported widely, a most impressive production run for a between-the-wars type.

Information relating to the Breguet 19s acquired by Bolivia has varied widely over the years. One very prestigious European aviation journal confidently reported at one point that not fewer than 15 were used by Bolivia. A close examination of all available evidence, however, suggests that a grand total of not more than seven were acquired, of two basic types, between 1924 and 1929.

The earliest recorded notation of a Breguet XIX in Bolivia was in August 1924, when one was on hand. Another was contracted for around 6 August 1925, and two more on 21 October 1926. The first three of these apparently all had 450hp Lorraine engines while the last one, and three subsequent aircraft, all had 500hp Hispano-Suiza engines.

At this time, Bolivian service aircraft carried names on the rear fuselage (rather than serial numbers, several of the Morane-Saulnier trainers being apparently the sole exceptions to this practice). The name given the first Breguet XIX is unknown (although it may have been "Oruro"). However, the next two aircraft were christened "Potosi" and "La Paz," and "Potosi" became the first aircraft to fly over La Paz at night, when this was carried out on 19 December 1926. Earlier, "La Paz" and "Oruro" made a highly publicized 'raid' on 10 November 1926 around the major cities of the Republic, Oruro, Potosi, Sucre and Cochabamba, returning successfully to La Paz. The famous French flyers Costes and LeBrix visited Bolivia during their epic journey through South America on 21 December 1927, and were escorted to the capital by at least one Breguet XIX ("Potosi").

An early view of a natural metal-doped Bolivian Breguet 19 A2 named "Batallon Colorados" with the standard script in use at the time instead of individual serial numbers. Just visible beneath the name is the small script "C. de P. y D. M." which denoted the rather awkward name of the service at the time (Ramiro Molina Alanes).

A popular subscription campaign was started on 29 October 1927 by the Committee of Propaganda and National Defense to acquire additional Breguet 19s for the Army, in the face of the growing tensions in the Chaco region. This fund subsequently raised sufficient funds to acquire three additional Breguet 19s and these were christened formally, shortly after arrival, on Sunday, 15 September 1929, some considerable time having been required to raise the requisite funds! The popular subscription was supplemented by a single 500hp (probably in actuality a 520hp Hispano-Suiza engined) Breguet 19 paid for by the Bolivian tin industrialist, Simon I. Patino, and this aircraft left Le Havre, France, on delivery to Bolivia on 3 October 1929, the last Breguet 19 acquired by Bolivia.

Press reports of the period, however, seem to rather cloud the issue of exactly how many Breguet 19s went to Bolivia, and when. A French periodical, "Le Document Aeronautique," for the period April 1926 to March 1927, quotes six Breguet 19s for Bolivia during that period, which seems at variance with the verified deliveries. A 1926 German publication seems to confirm this, "Autoflug-Anful" stating that six Breguet aircraft were being accompanied to Bolivia by a French mechanic sent by the manufacturer named Pierre Manut. Finally, an in-house Breguet publication dated January-February 1927 reported that a company pilot, Captain Henri Lemaitre, had returned to France after successfully delivering three Breguet XIXs to Bolivia, two of which had 450hp Lorraine engines and one the 520hp Hispano-Suiza engine. One of these was definitely "Potosi." During his stay, Captain Lemaitre apparently entertained the populace of La Paz with some very brazen aerobatics over La Paz in one of the aircraft.

Captain Lemaitre returned to Bolivia again, in the company of a mechanic named Faucher, apparently incident to the delivery of the last four aircraft in 1929, and aided the Bolivians in repairing several of the aircraft which had been idle for some time due to mechanical problems beyond the abilities of the Bolivians.

The Breguet 19s acquired in 1927 and 1929 were considerably improved over earlier aircraft, including a general strengthening of the structure to a gross weight of 3,500kg. Other improvements, not readily apparent in a superficial examination of the variants involved, included more effective ailerons (one of the chief criticisms of early Breguet XIX variants was that they were somewhat 'heavy' on the controls) which improved responsiveness markedly

Rarely photographed together, two of the Bolivian standard types show very different marking styles. The Fokker C Vb on the left appears natural metal/doped metal overall except the standard national markings, whereas the 450hp Lorraine powered Breguet 19 A2 (named "Potosi" in block letters on the rear fuselage) is a dark color (probably the French green in wide-spread use at the time) with an unusual dark vertical stripe at the leading edge of the rudder (Hagedorn Collection).

and a strengthened fireproof engine bulkhead. Fuel capacity was also improved to a potential maximum of 440 liters (two tanks, one of 320 liters and the other 120, which had the remarkable feature of being enclosed in 'fireproof' coverings and of being jettisonable in flight). Additionally, one of the bomb racks could be replaced fairly quickly by a third 400 liter auxiliary fuel tank, also protected and droppable in flight, which would give a potential fuel capacity of 840 liters. Other improvements included such detail comforts as a small windscreen for the observer, a bomb sight beneath the pilots seat, and provision for a fourth 100kg bomb-rack beneath the wings. With the 450hp Lorraine engine, Bolivia's Breguet XIXs were capable of a top speed of 124mph while the extra 70hp of the 520hp Hispano-Suiza engine in the Breguet 19 A-2 increased this to only about 136mph - certainly no speedster.

At least two of Bolivia's Breguets (including "Potosi" and probably "La Paz") were painted in the typical French over-all green of the period, with the standard Bolivian national insignia in the four wing positions only, and the national tri-colors on the upper portion of the rudder. The individual aircraft name on these was presented in large block letters on the fuselage sides. However, in the last (and majority) of the aircraft acquired, the machines were apparently doped aluminum overall and the individual aircraft names were in a flowing script on either side of the fuselage, atop the acronym "C. de P. y D. M." (which translated to Cuerpo de Pilotos y Departamento Militar) in small script, all in black. These aircraft included "Boquerón," "Batallon Colorados," and "Ingavi." One other Bolivian Breguet 19 is believed to have been named "Uncia," and the final example may have been "Oruro II." National insignia was carried in four wing positions only, while the entire rudder was used to present the national colors.

During the most serious of the border incidents with Paraguay during the period 1928-1932, the Breguet 19s were certainly the most important aircraft in the Bolivian inventory and, when joined by the new Vickers "Vespa IIIs" and older surviving Fokker C.Vbs, presented the nation with an apparently efficacious military capability.

Little is known in detail about the early Bolivian employment of its newer Breguet 19s, aside from the fact that a number of long-range reconnaissance missions were flown and one, none other than "Potosi," was employed on an extremely extended mission (supported by a former LAB Junkers F 13, "Illimani") to make an electrifying bombing raid on Paraguayan positions out of Puerto Suarez, on or around 18 January 1929 - very possibly the first aerial action

Close-up view of one of the Bolivian Breguet 19 A2s, "Ingavi," in natural/doped metal overall showing the twin 7.7mm guns in the rear cockpit and presentation of the service title script. Breguet 19s carried out some of the earliest raids of the conflict. (Hagedorn Collection).

Another, closer view of Breguet 19 A2 "Batallon Colorados." The Breguet 19 was a much larger aircraft than commonly thought, and suffered from its narrow track undercarriage at the rugged El Alto aerodrome (Hagedorn Collection).

of the conflict. It is of interest to note that this mission was flown by the French "advisor" Major Henri Lemaitre and his mechanic, Ernest Foucher, a fact known to the U.S. embassy at the time but seldom mentioned in Bolivian accounts of the war.

By the time actual full-scale hostilities had commenced, only three Breguet 19s remained combat-worthy, the rather narrow track of the Breguet 19s undercarriage having accounted for no end of disastrous ground-loops on the harsh landing field at "El Alto." These three were all sent to Villa Montes and Muñoz and, in October 1932, one of these engaged in a rather unlikely "dog fight" with a Paraguayan Potez 25. U.S. Army Air Corps Lt. A. Y. Smith, mentioned earlier in this account, who was traveling in South America as of 20 January 1932 reported that the Bolivians had "...three Breguets at El Alto," which would seem to indicate that the aircraft in the Chaco were not the only surviving examples by that date. A U.S. Intelligence Report dated 23 May 1933 reported that Bolivia had "two 520hp Hispano-Suiza Breguets at La Paz." Remarkably, all three of those which actually saw service in the war survived the rigors of the Chaco to be returned to La Paz in 1934, where they were apparently shunted off into a corner and, due to advanced age, lack of spares and general disinterest, were left to rot.

Bolivia officially credits the following pilots as having crewed Breguet 19s as part of the Escuadrilla de Reconocimiento during the war:

Cap. Eliodoro Nery
Cap. Leónidas Rojas
Tte. Arturo Valle

A third natural metal/doped Bolivian Breguet 19 A2 was "Boquerón" and the rudder colors appear to be considerably faded, and the tail skid seriously eroded! (Hagedorn Collection).

FOKKER C-Vb

Most reports and histories of aviation in Bolivia invariably quote the angular Fokker C-Vs acquired by Bolivia as C-Vc variants. However, recently it has been learned that the six (some sources say five) aircraft in question were, without doubt, actually C-Vb variants, and had manufacturers serial numbers 4874 to 4878, one of these numbers apparently used twice to produce the sixth aircraft, a not unknown Fokker practice at the time.

These aircraft, ordered 12 January 1925, were the first "standard" combat aircraft type acquired by Bolivia, and were purchased at the suggestion of Cpt. Raúl Vargas Guzmán, Technical Director of the Escuela Militar de Aviación, who apparently traveled to Holland to oversee the construction and delivery of the aircraft.

Both German and U.S. Diplomatic reports (dated 24 January 1927 and 12 December 1927, respectively) confirm that six Fokker C-Vs were, by those dates, actually on hand in Bolivia although an earlier U.S. report, dated 11 September 1926, only cited five. These aircraft had 520hp Hispano-Suiza engines, as did some of the Breguet 19s acquired by Bolivia, in an admirable attempt to standardize on one high-performance engine type.

The Fokker C-Vs have really never received proper credit for the contributions they made to the training of cadre during the years immediately prior to the onset of open hostilities with Paraguay. A number of long-distance, "good-will" flights were made with these aircraft to neighboring nations, and proved useful in building confidence amongst Bolivia's aircrews in their ability to navigate over unfamiliar terrain. One of the first of these was a much heralded flight commencing on 6 October 1928 to Argentina, the aircraft used being named appropriately "Bolivia" at the time and being crewed by Maj. Alfredo Santalla Esprella and Cap. Lucio Luizaga. The trip also included several stops in Uruguay on the return circuit.

Another flight, the next month, was attempted to Rio de Janeiro, Brazil via Santa Cruz, Puerto Suárez, Tres Lagoas, Baurú and Sao Paulo, crewed by Cap. Jorge Jordán Mercado (who became one of the heroes of the Chaco air war) and Sgto. Julio Loayza. Unfortunately, the aircraft (named "Mariscal Sucre") suffered engine failure on the leg between Tres Lagoas and Baurú, Brazil and the flight was terminated and the aircraft lost. However, the crew were still

Montage of views showing the Bolivian Fokker C Vbs in service. The center view shows the single example believed engaged early in the conflict in the Chaco region. In the early years of use, the Bolivian C Vs did not have individual serial numbers, being named after Bolivian victories and hero's, like the Breguet 19 A2s. (via George von Roitberg).

A clean view of two Bolivian Fokker C Vbs at El Alto about 1928. Unlike the Breguet 19s in service at the same time, the Fokkers had fuselage roundels but no obvious individual names as yet. (Antonio Sapienza Collection).

feted as heroes, as this route was considered especially audacious, being over largely uncharted territory.

One of the Fokker C-Vbs accompanied the Breguet 19 "Potosi" on its "reprisal" raid on Paraguayan installations in the Chaco on or around 18 January 1929, although the identity of this aircraft is not known. Crewed by pilot Sub.Tte. Faustino Rico Toro and observer Maj. Alfredo Santalla, this aircraft is noted to have had the uppermost national color on its rudder (red) carried all the way forward to the leading edge of the vertical fin, an unusual practice at the time.

The Bolivian C-Vbs colors are not known with any certainty, although they appear in available photos to have been doped aluminum overall in most cases, aside from standard national markings in four wing positions, either side of the rear fuselage (unlike known Breguet 19s) and, on most of the aircraft, the national tricolors on the rudder area only. Three are known to have been named: "Bolivia," "Mariscal Sucre," (believed actually worn as "Mcal. Sucre") and "Sgto. Max Paredes," while two others simply had Roman numerals "I" and "II" (although this is sometimes shown in records as "Fokker I" and "Fokker II" it was not presented on the aircraft in this manner).

By the time open hostilities erupted in 1932, only two of the Fokker C-Vbs remained on strength, and both were at La Paz, where they were largely ignored. The fate of the others in the interim is unknown. However, this wasn't the end of the Bolivian Fokker C-Vb story.

Cliff Travis of the Curtiss-Wright team in La Paz, reported as late as 13 September 1933 that "...the Bolivians have two old Fokker D-7s [sic] here that have been lying semi-derelict in the hangar for, reportedly, seven years come December more-or-less and they are going to have Don [of the Curtiss-Wright technical team] weld them up and get them on the job again. I have tried to talk them out of it on the grounds that the tubing is bound to be rusty and eaten inside and that they will turn out to be coffins for someone, but they have just received new wings and engines for them....which was some of [Raúl] Vargas Guzmáns' doings last year before they got wise to him and kicked him out of the service." Indeed, a U.S. Intelligence Report dated 23 May 1933 had reflected, somewhat inaccurately, that Bolivia had as of that time "one Fokker 550hp Hispano-Suiza acquired 1929(?)...and three 600hp Hispano-Suiza Fokkers acquired in 1933" which, while somewhat inaccurate in detail, probably accurately reflects that at least four Fokker C-Vb airframes still existed by that date at La Paz.

The subsequent fate of these aircraft is not reflected in any Bolivian history of the war, although JANE'S All the Worlds Aircraft still reflected the type as being in service in Bolivia into its 1939 edition! Intelligence reports on Bolivian air strength, however, do not show any Fokker aircraft later than 1933.

CURTISS D-12 "FALCON"

Ironically, Bolivia's powerful "Cyclone Falcons" were not the only variants of this basic classic design to see service during the Chaco War.

It will be recalled that Curtiss had established, with the cooperation and investment of the Chilean Government, a "factory" near Los Cerrillos, Santiago, Chile, where a series of Curtiss P-1A, P-1B "Hawk" and Curtiss D-12 powered "Falcons" were, essentially, assembled from knock-down component parts supplied by Curtiss.

For reasons that are beyond the scope of this account, this undertaking floundered as the depression advanced, and the Chilean Government essentially reneged on part of their contractual agreement for the supply of "Falcons" and a few "Hawks" for the Chilean Air Force. At least 10 and probably 11 D-12 powered "Falcons" remained unsold yet complete, and Curtiss was left to its own devices to attempt to recoup their investment in these aircraft and, additionally, get them out of the country!

By 1932, a bitter civil war was on-going in neighboring Brazil, and one of the factions, the Constitutionalistas, managed to purchase nine of these and arrange for them to be ferried to Brazil across Paraguay and part of Argentina, a most intrepid undertaking at the time and the full account of which is little short of astonishing.

Miraculously, seven of the aircraft actually reached Brazil. However, one, flown by an alleged "Argentine civil pilot" named Guillermo Hillcoat (but more likely a British subject and soldier-of-fortune, William Hillcoat), as recounted in the next chapter under the story of the Paraguayan Wibault Type 73 C-1's, was (according to the ofte-repeated account) "intercepted and forced to land" at Concepción on 24 August 1932. This aircraft was promptly incorporated into the strength of the Arma Aerea and performed a number of valuable reconnaissance missions on behalf of Paraguay.

In actuality, however, the aircraft was not "intercepted" at all but was, rather, the price of right-of-passage for the remaining aircraft enroute to Brazil, as they had to refuel in safety somewhere enroute. Contrary to previous accounts, this aircraft was, at the time of hand-over, not painted olive green overall, as were the others, but was otherwise completely devoid of markings, initially.

Although utilized primarily as a high-speed communications and staff transport (armed with a pair of 7.7mm guns in the rear gunners position, apparently borrowed from a Potez 25), the "Falcon" was given s/n 17 and, amongst many other duties, was used as an ad hoc "Air Force One," transporting the President of the Republic on at least one occasion, while also delivering vital medicines and mail. At least one reconnaissance flight, piloted by Maj. José Atilio Migone, was made in this aircraft on 26 April 1934, escorted by two Fiat CR.20bis fighters - a decidedly unlikely combination! This mission was to bring back vital information on the campaign in the vicinity of Camino "J" in the Chaco.

Incredibly, in spite of what must have been daunting maintenance challenges, the "Falcon" survived until 1943 and saw extensive use, and was finally withdrawn only after it was completely worn out. It has been further reported that this aircraft was fitted with some sort of a special canopy late in its service career to provide a greater degree of comfort for VIP passengers, but this has not been confirmed. The aircraft was apparently painted or doped aluminum or white sometime after acquisition.

Although neither Bolivia nor Paraguay acquired Curtiss D-12 powered Curtiss "Hawk" fighters, the aircraft pictured here was demonstrated and/or offered to both powers! Immaculate in overall doped natural metal, save the name "Curtiss Hawk" on the rudder, the aircraft was very ably demonstrated by none other than Jimmy Doolittle (left) Oscar Herrera and, at right, none other than C.W. Webster himself, who had perhaps more to do with equipping the Bolivian air arm during the war than any other single person. (Hagedorn Collection).

MACCHI M.18A.R.

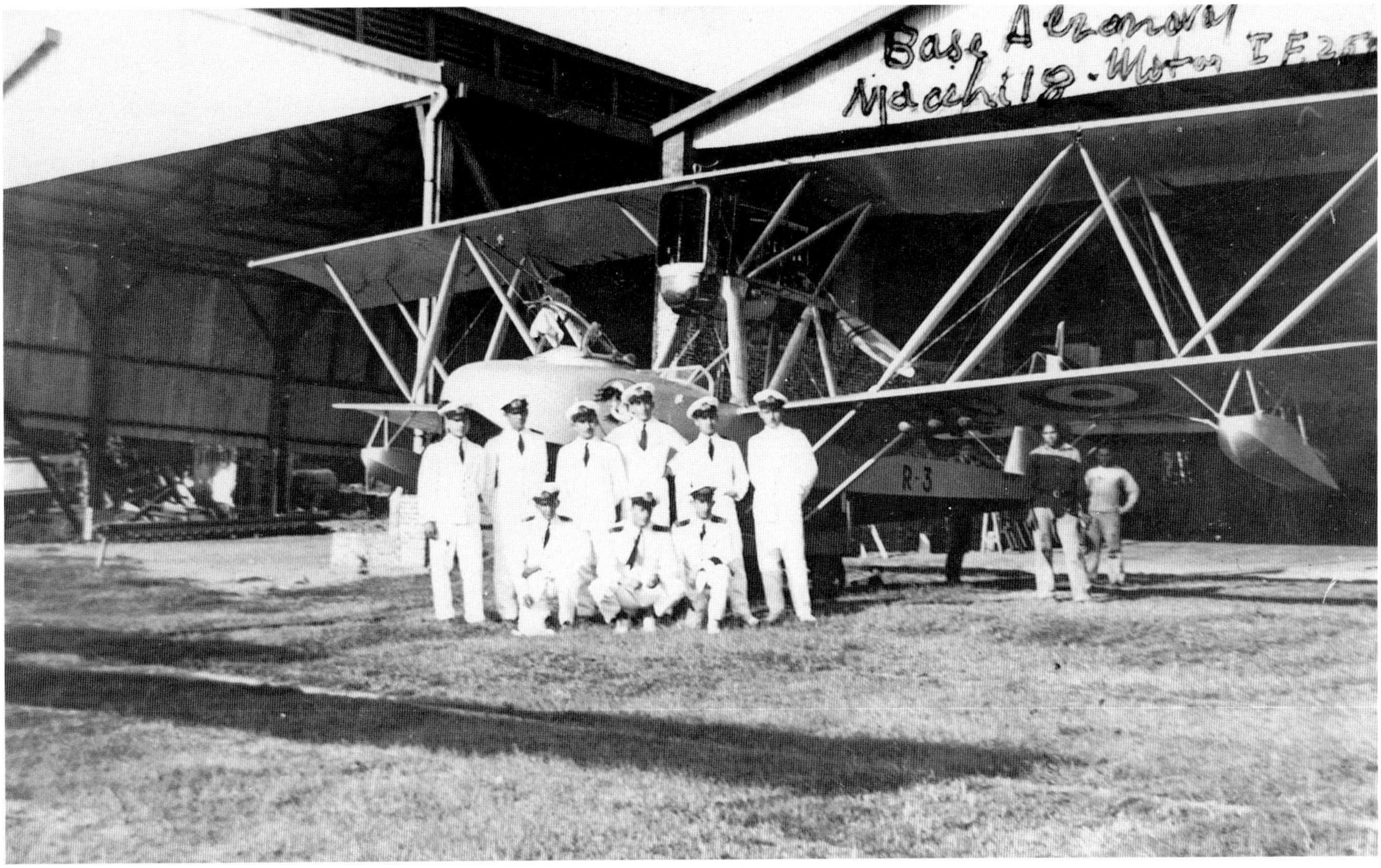

Complete with forward flexible gun and under-wing ordnance, this seldom remembered Paraguayan Naval Macchi M.18A.R. flying boat, R-3, is shown with its formally attired crews at its Asunción river base (Courtesy Roberto Gentilli).

Seldom recalled in discussions of the aircraft that saw action during the war were the two Macchi M.18A.R. reconnaissance flying boats acquired some time in either late 1932 or early 1933 for use by the Escuela de Aviación Aeronaval del Paraguay.

Derived from a basic 1920 Macchi design, the M.18A.R. (for "ali ripiegabili" or 'folding wing') featured a 250hp six-cylinder Isotta-Fraschini engine driving a pusher two-bladed prop, which could ideally deliver up to 114mph (184km/h) maximum speed. Maximum range was dependent on the crew and ordnance arrangement, which consisted of a single 7.7mm machine gun on a ring mount in the bows and four bombs of up to about 110 pounds each, two under each inner lower wing on racks.

The exact circumstances of Paraguay's acquisition of these two flying boats is unknown. Unlike the other flying boats operated by this small service, however, the M.18s were intended from the out-

The other combatant Paraguayan Navy Macchi M.18A.R., R-5, tied up along the bank at the Bahia Negra Naval Air Base. These aircraft performed valuable service and have received scant attention since the war (Capt. J. R. Ocampos/Aviación Naval Paraguaya).

Exceptionally rare in-flight view of Paraguayan Macchi M.18A.R. right after a bombing mission over Bolivian positions at the end of 1933. (Archivo Anibal Ferreira).

Head-on view showing the distinctive strut system of the Paraguayan Macchi M.18A.R., in this case serial R-5 at Sajonia Naval Air Station circa 1934 (Archivo Cnel Av. (S.R). A. Pasmor).

Paraguayan Naval Aviation crews aboard one of the Macchi M.18A.R. flying boats at Bahia Negra in 1934 showing the special insignia carried on several Naval aircraft. The forward gun seems to have a rather long barrel and the type is not identified (Courtesy Aviación Naval Paraguaya).

Another view of the bow turret of a Paraguayan Naval Macchi M.18A.R., in this case mounting twin Madsen 7.7mm guns. The officer at the guns is Lt.(Res). Job von Zastrow and the photo was taken at Bahia Negra in 1934 (Aviación Naval Paraguaya).

Macchi M.18A.R. s/n R-3 is seen here actually departing from its Rio Paraguay base at Bahia Negra on a combat mission in 1934. (Museo de Historia Militar via A. L. Sapienza).

Ground crews of the Paraguayan Navy inspect the underwing starboard bomb rack of a Macchi M.18A.R. at Bahia Negra in 1934. The special insignia carried on both M.18s is visible in the background (Aviación Naval Paraguaya).

set as combatant aircraft. They have been variously described over the years, which has led to some confusion. The otherwise excellent book Cincuenta Años de Aeronautica en el Paraguay, for example, cited the aircraft as "Nieuport-Macchi M-19's" with 160hp Fraccini engines (which, unfortunately, is both completely in error and very mis-leading), while other reports cite the duo as M.16's with 250hp engines or 200hp engines!

Issued Paraguayan Naval serials R-3 and R-5 (there was no R-2 or R-4), both aircraft were still airworthy and flown regularly as late as August 1934. R-5 suffered an accident some time in that same month, however and, by February 1935, both aircraft were inoperable and under repair, probably following the intensive operations in which they had been engaged. They were returned to service by May 1935, however but R-3 was completely destroyed at the hands of Midshipman Fretes Yodice in August 1935. R-5 soldiered on, however, but also suffered an in-flight fire while piloted by Tte. Martino following engine failure but the young officer managed to make a forced landing (on dry ground) and thus saved the aircraft from certain destruction.

R-5 in fact survived the war and was still reflected in the Order of Battle for the Paraguayan Navy as late as 1 July 1943, although it was partially dismantled by that date. It was still on strength, incredibly, as of 31 October 1947, although listed as "out of commission," thus earning the title of surely the last Macchi M.18 on active service anywhere in the world.

Like its transport sister aircraft the Savoia S.59, the two M.18s were apparently painted aluminum dope overall, except the planing bottom of the hull, which was black. The Paraguayan national colors were carried on the rudder and roundels in all four wing positions. The individual serials were on either side of the rear fuselage in approximate 12 inch black characters and both aircraft are known to have carried some sort of special insignia on the port bows (at least) just under the gun position, the exact character of which is not known with any certainty.

DE HAVILLAND/ADC D.H.9

Bolivia acquired at least two de Havilland (Airco) D.H.9s, one of which was operated on floats at one point. Here, AM-1, named "Coronel Salazar" is shown in what appears to be a Chaco-like environment, although there is no official record of the type having operated there (Courtesy G. A. Jenks).

Very little is known about the (at least) two de Havilland D.H.9s equipped with floats (initially) and acquired by Bolivia from the Aircraft Disposal Company (ADC) in England sometime in 1925.

These were apparently erected and test flown at Riberalta on the Rio Beni by the English pilot J. R. "Joe" King some time in 1925. Reputed by some sources to have been operated as "civil" aircraft, a photo of one of them in JANE'S for 1928 clearly shows the aircraft bearing Bolivian national insignia, including roundels on the fuselage, rudder stripes and the serial code AM-1. The other, reportedly AM-2, was similarly equipped with the 240hp "Puma" engine and was alleged to have been flown by another English pilot, Geoffrey Clapham until some time into 1930, when Clapham (also given as "Chapham") reputedly joined the Bolivian military flying school.

Yet another source claims the two D.H.9s were U.S. "surplus," though of course this is most unlikely.

One of the aircraft made a flight of note in 1927 of some 520km from Riberalta, which is in the far north of the Republic, to Puerto Maldonado to the west in neighboring Peru. If indeed these aircraft survived into the Chaco War period, they probably did so as Bolivia's sole aviation "presence" in the north of the Republic, far from the scene of any fighting, and thus are only coincidentally germane to this account.

However, a photo of AM-1, bearing the name "Coronel Salazar" also shows the aircraft on a conventional wheel undercarriage and in a Chaco-like environment. Therefore, the possibility that at least one of these aircraft may have survived into the early conflict period cannot be ruled out.

ANSALDO SVA-5 AND SVA-10

Surely the oldest aircraft to reach airworthy status - although just barely - into the Chaco War period were single examples of the Ansaldo SVA-5 (s/n 2) and SVA-10 (s/n 1) shown here alongside the fuselage of a S.A.M.L. A.3 used as a "hopper." (Prof. Carlos Pusineri).

Incredibly, two of the Ansaldo aircraft acquired during Paraguay's Civil War of 1922 remained in service with the Arma Aerea as late as 1932 and possibly beyond, and thus must be considered Chaco War participants, although only in training and support roles.

These aircraft were s/n 1 (an SVA-10) and s/n 2 (an SVA-5). Their ultimate fate is unknown.

3

Fighters and Fighter-Bombers

CURTISS MODEL 35A "HAWK II" AND MODEL 65 "SEA HAWK II"

The beautiful Curtiss "Hawk II" fighter-bomber, although acquired in comparatively small lots by Bolivia, etched an indelible legacy in the Chaco air war out of all proportion to its numbers. Here, s/n 40, the fourth example received circa August 1933, is shown in full markings in flight over the Chaco, complete with belly auxiliary slip fuel tank (Ramiro Molina Alanes).

Arguably the best all-round aircraft of any type to see action during the Chaco War, Bolivia's classic "Hawk II" and "Sea Hawk II" fighter biplanes, with their rakish spatted undercarriage and evocative lines came to symbolize this little known air-war in many minds.

The total number of "Hawks" acquired by Bolivia seems out of all proportion to the number of actions in which they are known to have participated. They seemed to be everywhere but, in fact, of the total of eight "Hawk IIs" and three "Sea Hawk IIs" ordered, only nine are known to have actually reached Bolivia before the U.S. arms embargo stopped deliveries.

Bolivia's first introduction to the Curtiss "Hawk" family had come, as with the "Falcon," at the hands of none other than Jimmie Doolittle, who brought an essentially stock D-12 powered P-1A "Hawk" with him to Bolivia for demonstrations in February and March 1928, and during which both the "Hawk" and "Falcon" were pitted in mock combat against the comparatively ungainly Bolivian Breguet 19s. This demonstration left a lasting impression on the Bolivians who, at the time, were under the guidance and advice of a mixed European advisory group.

One aspect of the Doolittle demonstrations that has not been previously disclosed, however, is the fact that the "Hawk" was initially shipped, along with a pair of pontoons, to Lake Titicaca, where it was hoped that the Curtiss team could interest Bolivia in launching a "Naval Air Service" based on a batch of float-equipped "Hawks." Unfortunately for posterity, Doolittle, despite his best efforts, couldn't coach the "Hawk" off the surface of the worlds highest lake, and the delegation continued on to La Paz.

Reports that Bolivia acquired as many as four "Hawk" 1A (P-1) D-12 powered fighters is completely without foundation, although there was at least some interest on the part of Bolivia in acquiring

"Hawk II" serial 32 receiving a maximum load of four bombs still looks factory-fresh, dating this photo sometime near January or February 1933 (Ramiro Molina Alanes).

aircraft of a very similar type then being "built" in the abortive Curtiss "factory" in nearby Santiago, Chile.

To gain a clear understanding of the utilization of "Hawks" during the Chaco War, it may prove instructive to examine the actual delivery sequence of the aircraft received. They were as follows:

Model 35A "Hawk II" c/n 11659/H23 officially delivered 19 Dec 32

Model 35A "Hawk II" c/n 11660/H24 officially delivered 19 Dec 32

Model 35A "Hawk II" c/n 11661/H25 officially delivered 19 Dec 32

Model 35A "Hawk II" c/n 11662/H26 officially delivered 19 Dec 32

Model 35A "Hawk II" c/n 11768/H64 officially delivered 3 Aug 33

Model 35A "Hawk II" c/n 11769/H65 officially delivered 4 Aug 33

Model 65 "Sea Hawk II" c/n 11824/SH27 offic. delivered 18 Jun 34

Model 35A "Hawk II" c/n 11844/H82 officially delivered 18 Jul 34

Model 35A "Hawk II" c/n 11845/H83 officially delivered 18 Jul 34

Two other Model 65 "Sea Hawk IIs", c/n 11825/SH28 and 11826/SH29 were to have been delivered 18 and 27 June 1934 respectively, but were apparently seized by customs officials before they could reach Bolivia. Their fate is unclear, but probably ended

A good look at the underwing bomb racks of one of the Bolivian Curtiss "Hawk II" biplane fighter-bombers. Lt. Claudio Lopez is astride the left wheel spat while the other pilot is unidentified (Ramiro Molina Alanes).

Believed to be the first three Curtiss "Hawk IIs" received by Bolivia in December 1932 (s/n 30, 32 and 34), this well-maintained trio is seen at Ballivian circa 1934. It is worth noting that the central green "dot" in the fuselage roundel of each aircraft appears to have not been painted on - or completely faded! (Ramiro Molina Alanes).

up in Peruvian service. Finally, a "Hawk II" often cited as being exported to Chile (c/n 11767/H46) circa 2 March 1935 seems in fact to have been destined for Bolivia and, according to at least one well-known Chilean historian, may have actually been delivered around the end of the war. However, this cannot be established with any certainty from available records. It should be noted that the "Sea Hawk IIs" were so named because they were specifically rigged to be mounted on either wheel or twin-float undercarriage, which indicates that Bolivia at least entertained the notion of operating several of the aircraft from the lakes and rivers of the Chaco. However, there is absolutely no indication that any such activities ever took place, and all of the Bolivian "Hawks" were operated as land planes.

Thus it can be seen that many of the initial battles of the war had been fought before the first "Hawks" could be brought to the front. Indeed, "Cliff" Travis, the Curtiss representative in La Paz, reported on 24 March 1933 that of the four "Hawks" which had arrived by that time, only two had been erected so far and, of these, "only one is in the Chaco, and seems to be all in one piece as of yet," but the second aircraft, which "has never left La Paz yet," was still there because "they have no pilots available that they are willing to trust with the machine".

The first "Hawk" seems to have been engaged in the first Bolivian assault on Nanawa which commenced 2 January 1933, this almost certainly having been c/n 11659/H23, which received Bolivian serial "30" (the other Bolivian "Hawks" were serialed 32 - later repaired following an accident and for some reason re-serialed as 48, plus 34, 36, 38, 40, 42, 44, and 46 in a more-or-less straightforward sequence at intervals of two, unlike other Bolivian Curtiss acquisitions).

"Cliff" Travis recorded that he conducted a test on the second "Hawk" that had been erected at La Paz on 28 March 1933 and reported that, "with full military load, got into the air in 800 feet on a wet field. The best take-off on a dry field here is 490 feet. It is considered nothing short of a miracle up here," and, with this demonstration, further cemented in the previously European-oriented Bolivian cadre, the preeminence of the "Hawk" and Curtiss family of aircraft.

The Bolivian "Hawks" were, on the surface of it, actually fairly good value for money, as the "Hawk IIs" purchased under Curtiss Sales Order S.O.125 cost $12,950 each, while the "Sea Hawks" with their optional float equipment, on Sales Order S.O.131 were $14,015 each. However, this was the sales price to Curtiss-Wright Export Corporation, which of course was the broker. The actual retail price to Bolivia for the Hawks was not less than $25,000 each, although "spares will be provided at list price," according to "Cliff" Travis' report.

The delivery dates for the Bolivian "Hawks" are somewhat misleading, as these were apparently the dates that the aircraft actually reached the South American "ports of entry" for Bolivia, which alternated, depending on the mood of the respective governments, between Arica, Chile and Mollendo, Peru. Customs officials at these ports caused no end of problems for the Curtiss brokers and, when combined with the seemingly endless vacillating of the respective governments on whether or not to allow arms to continue to flow to Bolivia, resulted in considerable delays in the transshipments. By May 1933, the movement for a total embargo by the United States had been well advanced, and Curtiss-Wright Export Corp. had already had to commence resorting to various subterfuges in order to fulfill its previously obligated contracts. The nearly 100% mark-up on the cost of the "Hawks" for Bolivia was due, in no small part, to what "Cliff" Travis described as "the unusual selling expenses....which, all in all, should provide plenty of margin on these." The `unusual selling expenses' were, of course, bribes to Customs and contracting officials along the way.

It is interesting to note that at least one attempt was mounted, by none other than United Aircraft Corporation (the major U.S. competitor to the Curtiss-Wright empire) to acquire "Hawks" from the Curtiss Aeroplane Co. in Buffalo direct and, acting as broker, to sell them to Bolivia at around $23,000 each, thus beating Curtiss-Wright Export's own price! As United had its own sales organization in place in Peru at the time, this could have probably been easily arranged. However, C. W. Webster, President of Curtiss-Wright Export, prevailed upon Corporation management to block such a deal in the interests of his proven sales team. Around the middle of February 1933, the Curtiss-Wright Export team also made a half-hearted attempt to sell the unsold remaining Curtiss P-1A "Hawks" with D-12 engines still at the Curtiss "factory" in Santiago, Chile, to the Bolivians "at a certain price," but this initiative was apparently blocked by the Chilean Government, with whom Curtiss was having no end of problems regarding the whole "factory" affair.

By 6 November 1933, by which time six "Hawks" were in country, Owen Shannon, who replaced "Cliff" Travis as the Curtiss-Wright Export rep in La Paz, was able to report back to Webster that "the other business of more "Ospreys" and "Hawks" will come along as a matter of course, as they are absolutely sold on them from the President right down to every pilot....and orders will be placed just as soon as they can get exchange." One of the final "Hawks" received, circa 13 September 1933, however, arrived with severe damage to the stabilizer while another arrived with the wrong type of nut for the bolts attaching the prop to the crankshaft, necessitating the delay of delivery of these two aircraft badly needed at the front (the last aircraft was almost certainly the single "Sea Hawk II" actually delivered, which, in fact, only operated as a land plane).

Unbelievable as it may seem, even at this point in the war, there was considerable in-fighting amongst the Bolivian leadership over arms acquisitions and alternate sources. Colonel Bernardino Bilbao Rioja, Commander of the Bolivian air arm at the time, for instance, complained to the General Staff of a number of "shortcomings" which had surfaced regarding the Curtiss equipment which, according to the Curtiss representatives, was fed completely by the fact that Bilbao was predisposed towards European equipment and (possibly for financial reasons) the equipment offered by United Aircraft Corporation. In an unprecedented (and nearly mutinous) move, "Cliff" Travis reported that "they who have been flying our equipment went to the Comptroller and General Staff on their own initiative and made written statements regarding the planes. These statements were by no means vague and every one boosted our stock. As far as I have been able to find out, there was not a complaint made by anyone but Bilbao himself. Colonel Jordan took the matter up directly with Bilbao and General Kundt and stated bluntly that he had [by then] nearly 500 hours in the "Hawks" and more than 250 in "Ospreys" and that they were absolutely satisfactory in every respect." He allegedly went on to say that "When you birds begin to know something about flying, you will see why the Curtiss planes are the best that we have ever had!"

As will no doubt be obvious to the reader, the sleek "Hawks" were seldom employed strictly as "pursuit" aircraft in the conventional sense, as the opportunities for air-to-air combat were rare indeed. More often, the aircraft were utilized as fighter-bombers, in company with other aircraft types in mixed formations and, occasionally, on single aircraft sorties against targets of opportunity in dense ground action situations. While the aircraft was very popular amongst some of Bolivia's best pilots, it was hampered in actual operations in the Chaco region due to the fact that its 700hp "Cyclone" engine, which was supposed to be fueled with 87 octane aviation gas, was usually run on 70 octane fuel, with a resultant

"Special" insignia on Bolivian aircraft during the Chaco War are virtually unknown. However, this Curtiss "Hawk II" bears the "iron cross" style unit insignia of Grupo 1, just visible ahead of the officer in the middle (courtesy TCnel. Amalia Villa de la Tapia).

The rather spartan instrument panel of the Bolivian Curtiss "Hawk IIs" which, even at that, was more than they were accustomed to! (Curtiss).

effect on performance and reliability. Thus, performance figures for the Bolivian "Hawks" are somewhat misleading, although they are quoted here as a basis for comparison. With the "Cyclone" engine (variously rated, as the SR-1820F-2 at between 600 and 700hp), the "Hawk II" supplied to Bolivia was advertised as being capable of a maximum speed of 202mph (325.08km/h) with a constant speed prop, making it - even with fuel problems - without question the fastest aircraft of any type of the Chaco War, range (without a custom slipper auxiliary tank, some of which were apparently supplied to Bolivia) was 522 miles (840km) and service ceiling was given as 25,100 feet (7,650m). As far as can be determined, all Bolivian "Hawks" were delivered with two 7.7mm machine guns, although one weapon of this caliber and another of 12.7mm could be accommodated. Additionally, the "Hawk II" could mount four bombs, two on a rack under either lower wing, of up to 110 pounds each. However, the aircraft seems to have seldom carried this heavy load, more often being seen with but one bomb under each wing of varying weights.

Despite their small numbers and the debilitating performance handicaps due to low-grade fuels, the "Hawks" seemed to be everywhere from January 1933 on and, remarkably, only one is known to have been lost to enemy action. This, s/n "30" (the very first machine received) was lost near Macheriti on 26 December 1934 while flown by Tte. Carlos Lazo de la Vega. The circumstances of this action are necessarily rather obscure, however, and, while he is officially credited as having been brought down by ground fire during his ground attack mission against the Paraguayan Fortín Puesto Central, it has also been suggested that he was actually the victim of a bomb dropped by a Bolivian "Osprey" which was engaged in strafing the same target at the time.

So far as can be determined, the largest number of "Hawks" engaged in any single mission was on 24 January 1935 when three accompanied three Junkers K 43hs, two "Falcons," two "Ospreys" and a Vickers "Scout" on a strike mission from Villa Montes.

Bolivia's "Hawks" were all painted natural aluminum dope overall, with national insignia in the standard four wing positions, either side of the rear fuselage, and standard rudder stripes. The individual aircraft serial numbers, described earlier, were carried in about 12 inch black, block numerals on either side of the vertical fin. One "Hawk," that flown by Tte. Emilio Beltrán, one of Bolivia's best pilots, is known to have been referred to by him as "Sr. Hawk" but it is not clear if this was actually presented on the aircraft in any way. Another one or two (s/n 36 and 44) assigned to Grupo No.1 at Villa Montes, is known to have carried the only "unit" insignia of any Bolivian air unit of the war, in the form of a German style maltese cross with a stylized "G.1" in the center field. Grupo No.II, also based at Villa Montes at the time with (amongst other types) "Hawks" s/n 34 and 42, apparently did not adopt any special insig-

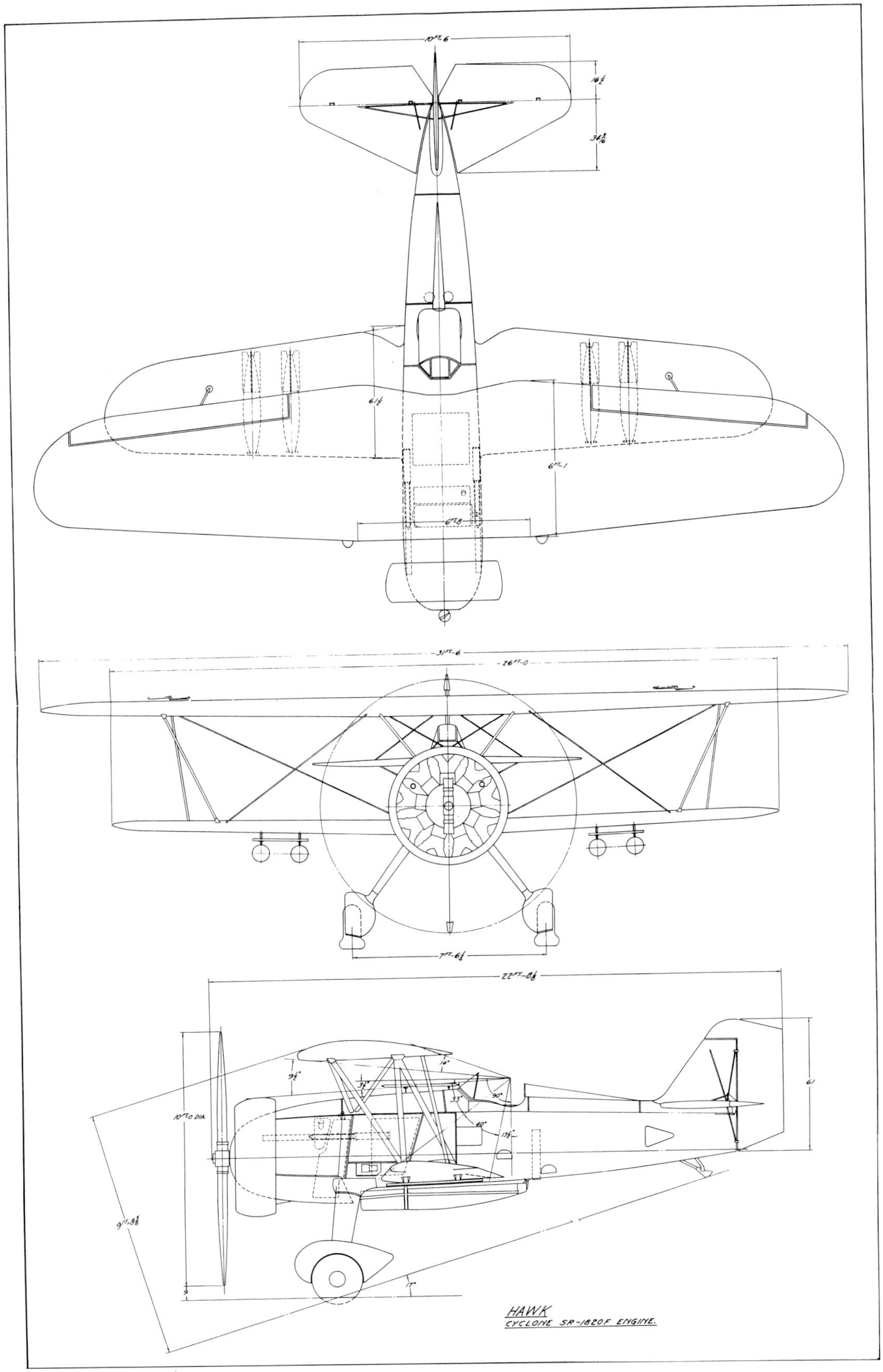

Three-view of the Curtiss "Hawk II" as supplied to Bolivia, complete with belly slip tank. The aircraft operated most often as a fighter-bomber (Curtiss).

nia so far as can be determined. Several of the "Hawks" were nominally part of the famed Punta de Alas squadron at one time or another. Another "Hawk," s/n 40, flown by Maj. Rafael Pabón, is reported to have been named "El Tigre-Hawk" but, again, it is not clear if this name was actually painted on the aircraft.

Mention of Major Pabón should not pass without mention of just how close Bolivia came to not enjoying the services of this outstanding pilot during the war. Rafael Pabón spent three years in the United States as an enlisted member of the U.S. Army Air Corps as a mechanic and had received his secondary education in La Paz at the American Institute, where part of the curriculum was a course in English. Following his entry into the USAAC, he had served as the personal mechanic for Major C. F. Woolsey, one of the U.S. pilots who later participated in the famous "Pan American Flight," killed in a collision in his Loening amphibian at Buenos Aires. Pabón had apparently saved nearly all of his enlisted pay in the USAAC and, on weekends, had taken flying lessons at the civilian Curtiss Flying School in San Antonio, Texas. Major Woolsey and other U.S. pilots who knew Pabón on the Pan American Flight team, commented that Pabón had, through this discipline, become "a very skilled pilot." Pabón had, following his USAAC Honorable Discharge (where he had attained the rank of Private First Class, Specialist 3rd Class), returned to Bolivia and petitioned for a commission in the Bolivian Army. This was resisted by the established Cuerpo de Aviación cadre which, it will be recalled, was entirely European trained and influenced at the time. Pabón thus set out to take his case to the people, and, in a series of speeches throughout the country (during which he reportedly wore the uniform of a USAAC officer!) pressed the flesh and generally promoted himself as worthy of a direct commission in the Cuerpo de Aviación. He went so far as to convince Cuerpo officials to "loan" him an aircraft circa 3 March 1927 (apparently one of the Caudron C97s, a type with which he was completely unfamiliar) so that he could show them ..."how a USAAC officer could fly." Flying on six consecutive mornings, he suffered an accident on landing on the sixth flight, and this event was seized upon by the established cadre to illustrate his "incompetence." His public "stumping" for a commission and accident in the precious C97 sparked a "very bitter" two-month controversy in Bolivia, during which the matter seemingly dominated the news in the country. The Commander of the Cuerpo de Aviación at the time, Teniente Coronel Génery Blacut (who, himself, was not a pilot, and who in fact had never even been aloft in an aircraft, even as a passenger) was, in the words of the U.S. Legation...."controlled entirely" by the Swiss and French Mission personnel. Pabón, for his part, alleged that the European instructors ....."only dashed about La Paz in one of the two Packard autos the government bought for them" and spent precious little time in the air themselves. None other than Ira Eaker, leader of the Pan American Flight (which passed through La Paz and members of which were questioned regarding Pabóns competence) also commented favorably on him and, as a result, no less than the President of the Republic himself finally tendered him a Commission in the Army as a Sub.Tte.....and promptly assigned him to the "La Loa" Infantry Regiment then in action in the Chaco. Fortunately for posterity, his father, a wealthy merchant in La Paz and related to many of the leading families in the capital, interceded again at this point and Pabón was finally transferred to the Aviation Corps - although he reportedly flew for a time as an Infantry Officer before finally being designated a Aviador.

The Bolivian investment in the "Hawks" resulted in one confirmed air-to-air "kill" for the type as a dedicated pursuit aircraft during the war. This was Paraguayan Potez 25 s/n 13, which was downed 11 December 1934 near Capirenda by Sub.Tte's Alberto Alarcón and Emilio Beltrán, both flying "Hawk IIs", a shared victory credit being awarded.

At least five of the "Hawks" survived the Chaco War (suggesting that several may have been lost in unrecorded training accidents) as five made a memorable aerobatic appearance in Lima, Peru on 23 September 1937 at the "III Conferencia Tecnica de Interamericana de Aviación" held there that month. By 15 March 1943, all five were still on hand, although they had been, prior to this report date, cited as "grounded," although one of these was lost on 5 April 1943 in a violent ground loop at the hands of Cap. Fernando Guillarte. Two of these were overhauled and returned to service, however, and remained on the Bolivian Air Order-of-Battle until at least 6 December 1946, when they were assigned (with one other, non-operable) to the Escuadron de Entrenamiento No.1 at "El Alto." Like the surviving Ospreys, Bolivia had requested Lend-Lease funds to obtain spares or repairs for three Hawks on 19 March 1945, but there is no evidence that this was ever approved or followed through with.

Bolivian alumni of the "Hawk" era are a select group, and in spite of their small number, the following pilots are known to have flown "Hawks" during the war:

Maj. Jorge Jordán
Cap. José P. Coello
Cap. Rafael Pabón
Cap. Luis Ernst
Cap. Leónidas Rojas
Cap. Juan Antonio Rivera
Tte. Alberto Paz Soldán
Tte. Emilio Beltrán
Tte. Alberto Alarcón
Tte. Fabián Monasterios
Tte. José Chacón
Tte. Luis Paredes

FIAT CR.20bis

The principal Paraguayan fighter type during the Chaco War was the Fiat CR.20bis, which was plagued by engine problems. Here, serial 11-7 is shown at Ñu-Guazu aerodrome in 1933, complete with unit "Los Indios" insignia and the chevron tri-color used on the wings in lieu of roundels (Archivo Crnel. Av. (R.S). A. Pasmor).

It didn't take the Paraguayan intelligence apparatus long to determine that Bolivia was launching a determined foreign arms acquisition program. This information, coupled with the disappointing performance of the Wibault fighters acquired through the equally ineffective French Mission, led Paraguay to seek a ready source for effective combatant aircraft to counter the expected Bolivian acquisitions.

Actual negotiations through the Italian Legation led the Paraguayan authorities to the Fiat CR.20bis, which had first been announced to the foreign press in March 1930. Ironically, Paraguay had, through a rather circuitous route, also been offered the Curtiss-Wright C14R "Osprey" at this juncture, which the Paraguayans judged "superior" in most regards to the CR.20bis at the time for their purposes but, apparently because delivery complications and funding requirements could not be met, this deal was allowed to pass into oblivion.

Incorporating improvements over previous models of the CR.20 series, the CR.20bis was a much strengthened aircraft, its main wing spars being made of seamless drawn duralumin tubing of parallel,

Note the "eyes" on the radiator of this Paraguayan Fiat CR.20bis at Campo Grande aerodrome near Asunción in 1935. Paraguay wrung the most from the CR.20, but its pilots frequently were frustrated with the aircraft (Antonio Sapienza Collection).

The remains of Fiat CR.20bis serial 11-1 near Isla Poí aerodrome after Tte. Walter Gwynn crashed and was killed there on 12 June 1933 (Antonio Sapienza Collection).

engine." Performance was adequate, if not exceptional, and depending on the horsepower rendered by the rather unreliable engine (International Rating of 410hp), the CR.20bis was listed at 155mph maximum speed (250km/hr, although Italian advertising had promised 172mph.), an endurance of approximately 2.30 hours at normal cruise and a ceiling of 24,608 feet (7,500m, although, again, Italian literature had suggested 28,000 feet.). The aircraft were supplied with two 7.7mm Vickers machine guns with 500 rounds per gun (although the CR.20bis was configured, if need be, to mount four such weapons, two in troughs on either side of the fuselage and the standard two on the top of the engine cowl), firing synchronized through the prop, just as with the Bolivian "Hawks." This armament proved something of a problem in itself, however, as it was found that the ammunition required for these guns was slightly different than that for the guns supplied on the Potez 25s and other aircraft.

Paraguay acquired just five Fiat CR.20bis fighters, and these did not arrive in Asunción until April 1933, by which time Bolivia had shocked the Paraguayan forces in the field by introducing the sleek Curtiss "Hawks." Interestingly enough, however, a Mr. Rogers of the U.S. Legation in Asunción claimed to have seen seven "new Fiats" at Campo Grande on or around 10 May 1933, and has led to speculation that in fact seven were acquired and two lost in early training accidents. This seems to be borne out by Adrian English' Armed Forces of Latin America, which states that "three Fiat CR.20s were written-off within days of delivery." In any event, however, shortly afterwards, following pilot selection and hasty conversion training, a new unit was formed to mount the new aircraft, for which the Paraguayan's harbored high hopes, the Undecima Escuadrilla de Caza "Los Indios", which was popularly known simply as "Los Indios" ("The Indians"), the aircraft receiving squadron/ serial codes 11-1, 11-3, 11-5, 11-7 and 11-9.

octagonal shape. This unusual structure was further braced with internal duralumin strips and rivets and built-up duralumin ribs were then secured to the spar by means of steel, rivetted "tangs." The main structure of the fuselage itself was entirely of steel tubing and the main fuel tank, aft of the engine in the lower forward fuselage and forward of the cockpit, was covered in a French made material called "Semape" which offered a form of bullet-proof protection, the CR.20bis being unique during the Chaco War in this regard. Additionally, the main tank could be jettisoned by the pilot, if need be - another sales feature apparently arrived at as a result of stories of World War One era pilots going down in flames behind fuel tanks ablaze. While the wings and part of the fuselage were fabric covered, just as on the Bolivian "Hawk IIs", otherwise the aircraft was all-metal.

Even as early as the date of the first U.S. intelligence report on the CR.20bis, however, it was reported that ..."difficulties are understood to have been encountered in cooling and lubrication of the

Two Fiat CR.20bis fighters of the Undecima Escuadrilla de Casa, Los Indios at Ñu-Guazu aerodrome in 1933. The tri-color chevron national insignia appeared on both upper and lower surfaces of the top wing and, it is believed, on the lower side of the lower wing only (Tte. Gonzalo Palau).

Extremely rare panorama view of the Paraguayan operating base in the Chaco at Isla Poí in 1933. From left to right can be seen a Potez 25, a Fiat CR.20bis, the elusive Curtiss D-12 "Falcon" and another CR.20bis. (Ministerio de Defensa Nacional).

Nine pilots from the former 1er Escuadrilla de Caza were detailed to man the new unit:

Cpt. Bernardino Caballero Alvarez
Cpt. Vladimir Porfenenko
Tte. (Navio) Victor Urbieta Rojas
Tte. Walter Gwynn
Tte. Tomás Rufinelli
Tte. Román García
Tte. Orlando Salerno Netto
Tte. Abelardo Bertoni
Tte.2° Abdón Alvarez Albert

The unit spent most of April and May 1933 conducting intensive familiarization training at Campo Grande near Asunción.

Once declared operational, all five aircraft departed by air from Campo Grande on 25 May 1933 enroute to the city of Concepción that day, and then continued their journey to Isla Poí base deep in the Chaco. All five arrived safely at Concepción, despite heavy fog over the city and, on the 26th, they took off to continue to Isla Poí. However, after only a few minutes enroute, Fiat 11-9 had to make a hasty forced landing near San Salvador due to mechanical problems and the aircraft, flown by Tte. Walter Gwynn, was completely destroyed in the process, although Gwynn survived. Another of the aircraft, s/n 11-7, also experienced mechanical problems, but returned safely to Concepción. Thus s/n's 11-1, 11-3 and 11-5 arrived at Isla Poí after a rather inauspicious start to their operational career.

So far as can be determined, the CR.20s first saw action on 11 June 1933, when four Bolivian "Ospreys" swept in to attack Isla Poí. S/n's 11-1, 11-3 and 11-5 scrambled immediately and moved to intercept the Bolivian aircraft. The "Osprey" crews were obvi-

Another lineup view of two of the Paraguayan CR.20bis fighters (courtesy Dr. Roberto Gentilli).

Excellent study of the empennage of one of Paraguays Fiat CR.20bis aircraft, manufacturers serial number 431, with a closeup of the operating unit insignia and Capt. P.A.M. Leandro Aponte, Director of the Paraguayan air effort in the Chaco, at Isla Poí in 1934. (Museo de Historia Militar via A. L. Sapienza).

The next to last Paraguayan Fiat CR.20bis, serial C-2 crashed at Campo Grande 19 October 1939 (the other aircraft survived to be dismantled in 1941) (Antonio Sapienza Collection).

ously surprised to encounter the Paraguayan fighter opposition and quickly jettisoned their bombs at random and made their escape.

This initial encounter apparently prompted a maximum effort response by the Bolivians, as, the next day, 12 June 1933, a force of five Bolivian "Ospreys," escorted by three "Hawks" and at least one venerable Vickers "Scout," attacked Isla Poí. The defenders had apparently received some advance warning of this force, as the three CR.20s had been scrambled and immediately engaged the "Hawks" - the first known encounter between fighters of both sides. S/n 11-1, flown by the now recovered Tte. Walter Gwynn, 11-3 at the hands of Cpt. Bernardino Caballero Alvarez and 11-5 with Tte. Tomás Rufinelli, Jr., were all airborne and anxious to try out their virtually brand-new aircraft against the legendary "Hawks." Tte. Rufinelli almost immediately experienced problems with his guns (reportedly just as he was about to down a Bolivian "Hawk") and, after a rather frantic dog fight, all of the Bolivian "Hawks" left the scene unscathed, apparently abandoning their escorted bombers.

One of the leading Paraguayan flyers, Capt. P.A.M. Leandro Aponte is shown in the cockpit of one of the Fiat CR.20bis fighters just before launching on a Potez 25 escort mission in 1933. (Museo de Historia Militar via A. L. Sapienza).

However, the Fiats quickly found the five "Ospreys," still attended by the single Vickers "Scout," about to plaster their base. The Fiats attacked the entire formation, which again quickly jettisoned their bomb loads and scattered. The hapless Vickers "Scout," piloted by Cpt. Luís Ernst, turned on Tte. Gwynn's s/n 11-1, and loosed several bursts. Since this Fiat did in fact crash near Isla Poí, killing the pilot, the Bolivians claimed it as an aerial victory. However, a Paraguayan technical commission, formed to investigate the incident, found no evidence of any bullet strikes on the aircraft or the body of the pilot. It was thought that Tte. Gwynn, who had not fully recovered from his earlier injuries, had passed out when returning to the base and crashed. However, the possibility that the Bolivian aircraft caused the Fiat to enter into an ill-advised maneuver in a defensive move cannot be ruled out.

Fortunately, by August 1933, s/n 11-7 had been repaired and delivered to Isla Poí and, together with 11-3 and 11-5, performed a large number of escort missions with (usually) Potez 25s over, amongst other contested locations, Platanillos, Saavedra, Corrales, Herrera, Samaklay, Zenteno, Falcón and Gondra.

Yet another of the prized Fiats, s/n 11-3, was unfortunately lost in another accident on 4 March 1934 at Isla Poí while being flown by Cpt. Bernardino Caballero Alvarez, who died instantly in his totaled aircraft. The 1 1er Escuadrilla de Caza was thus reduced to a total effective strength of but two aircraft within the space of 11 months of operations.

During August and September 1934, the surviving aircraft (serials 11-5 and 11-7) escorted Potez 25s in an endless series of bombing and reconnaissance missions over, amongst others, Strongest, Garrapatal, Picuiba, Algodonal, Carandaiti (given as Carandayty by the Paraguayans) and Ybamirante. During one of these, on 23 September 1934 over the Picuiba area, 11-5, flown by Tte. Tomás Rufinelli and 11-7 by Tte. Orlando Salerno Netto, while escorting Potez 25 A-2 s/n 5 (flown by Tte. Homero Duarte and Tte. Cándido Ríos, his gunner) sighted two Bolivian "Ospreys" laden with bombs.

Both Fiats dove to attack the lower "Ospreys" when one of them, to the surprise of the attackers, suddenly spotted the Fiats, jettisoned its bombs and maneuvered away to escape. The other "Osprey," apparently oblivious to all of this, continued on towards its target for a short while but, eventually, must have noticed the absence of its wing man (or the presence of the Fiats) and did likewise - not, however, before the Paraguayans got in some hits on the second "Osprey." Although allegedly severely damaged, the "Osprey" returned to its base, the Paraguayans claiming only a "probable" as a result of this encounter.

Both of the surviving Fiats spent the remainder of the war in yet further escort missions and survived to participate in the Victory Parade on 20 August 1935. The operating unit was then demobilized and inactivated and the aircraft were transferred to the 1er Escuadrilla de Caza, which eventually received Fiat CR.32s.

During the post-war period, the two surviving aircraft were re-serialed as C-1 (former 11-5) and C-2 (11-7). C-2 was subsequently destroyed in a flying accident on 19 October 1939, killing its pilot, Dr. Carlos Nogués, at Campo Grande. C-1 remained active until 1941 when it was finally dismantled due to age, wear and tear.

Paraguays Fiat CR.20s were painted, oddly, in an overall white color scheme (close to Humbrol HU034/FSN37875) with the exception of the forward half of the fuselage from the engine to the cockpit, which was aluminum (Humbrol HU056) when first acquired. However, it is not generally known that shortly before or after deployment to the field, they were painted/doped aluminum overall, perhaps because the white finish, while nice for parades, was bound to present maintenance problems in the rugged Chaco environment. The Paraguayan national colors were on the rudder with a yellow star on the middle (white) band. The unit insignia, an Indian (painted in dark brown, and not red as has been reported), was painted on both sides of the fuselage just aft of the cockpit and, unusually, the CR.20s did not carry the Paraguayan roundels on the wings but, rather, had red/white/blue stripes in four wing positions (some in a chevron-like presentation), probably in an effort to distinguish them, in the heat of combat, from the roundels of their Bolivian adversaries. Serial numbers were painted in small white numerals with black outlines on both sides of the fuselage, Italian style, under the horizontal stabilizers. There is some evidence that the two aircraft which survived the war were repainted natural aluminum overall and that roundels then replaced the wing stripes, but the old "Indios" unit insignia was retained, even though the unit with whom it was associated no longer existed.

Seldom photographed from the rear, one of the Paraguayan Fiat CR.20bis at Isla Poí in 1933. This aircraft does not have the unit insignia of the 11th Fighter Squadron on the fuselage (the only known instance) nor the diagonal stripes on its wings noted on all other examples. (Museo de Historia Militar via A. L. Sapienza).

VICKERS TYPE 143 "BOLIVIAN SCOUT"

Certainly one of the least attractive fighters of all time, the rather ungainly looking Vickers Type 143 "Bolivian Scouts" actually turned out, like their equally ugly bomber/recce hangar mates, the "Vespa IIIs", to have been a good investment for Bolivia.

Part of the total arms package arranged via the Dillon Read Loan to the country, it has always been stipulated that Bolivia acquired a total of but six "Bolivian Scouts" on Vickers Sales Order No.85-A at 30,433.12 Pounds Sterling. However, the "package" deal apparently included at least two or three spare fuselage sections and extensive spares of other categories, which would prove useful later in the service lives of these little-known aircraft.

First developed as the Type 123 "Hispano Scout" in 1926, through the Type 141 "Scout" of 1928, both types featured all-metal construction with fabric covered surfaces. Significantly, when pitted against Weybridge-built Wibault Scouts, the Vickers aircraft turned in a better speed and ceiling, which may have been a factor in Bolivia's choice of the next development in this series, the Type 143.

The Type 143 was based on the Type 141 but were fitted, instead, with 450hp Bristol "Jupiter VIA" air-cooled radial engines, not accidentally the same engine as mounted on the "Vespa IIIs" that were part of the same arms package. Besides the engine, the other major changes from the earlier Type 141 were the adoption of a split-axle undercarriage, a Fairey-Reed metal prop and additional streamlining of the fuselage, to match the characteristics of the radial engine.

A reversion of wing-tip hoop-skids (common on World War One era aircraft) was also specified, since the aircraft were expected to be flown from very rough South American fields in Bolivia, where, indeed, the hoops and strengthened undercarriage certainly prevented far more extensive damage to the Type 143s than would have otherwise have occurred in hard landings.

Another of the ungainly appearing trio of Vickers types acquired as a package by Bolivia pre-war, the Type 143 "Bolivian Scout" actually gave a very good account of itself. Here serial number 8 is seen at Villa Montes in 1932, fully combat ready (Ramiro Molina Alanes).

Taken shortly after delivery and acceptance, four Vickers Type 143 "Bolivian Scouts" on line with three "Vespa IIIs" in the distance at El Alto prior to the war. The Type 143 was very maneuverable, and gave the Bolivian pilots great confidence. (Courtesy E. B. Morgan).

Ordered by June 1929, the six "Bolivian Scouts" arrived at La Paz between 3 September and 17 December 1929, where they were assigned individual serial numbers 7, 8, 9, 10, 11 and 12.

Two of the Type 143s were lost in training accidents before full-scale hostilities commenced and, sometime after 1931, three of the surviving four aircraft were, for reasons unknown, reserialed as 54, 59 and 96 (s/n 8 keeping its original serial). It has been speculated that, in fact, these "new" serials were aircraft built up from the spares mentioned earlier.

Prior to the outbreak of full-scale war, the "Bolivian Scouts" had been rather intensively flown, and amongst their more noteworthy involvements were the escort (by s/n 7 and 10) of Brazilian Army Amiot 122 serial K-622, Conde de Caxias, which made a Goodwill flight to Bolivia on 12 October 1931. Cpt Rafael Pabón Cuevas, cited at length earlier in this account, also added a feather to his cap in 1931 when he flew one of the Type 143s to an altitude of 10,500m - a world record at the time.

The surviving four Type 143s saw a surprising amount of action during the ensuing conflict, during which two more of them were lost in accidents. One of these occurred at "El Alto" on 8 April 1932 when, while piloted by Sub.Tte. Faustino Rico Toro (an early hero of the war, having piloted the Fokker C-Vb that conducted one of the earliest bombing attacks on Paraguayan forces) one of the aircraft (probably s/n 9) crashed during aerobatics practice. During the battle for Boqueron (the first full pitched battle of the war), s/n 7 and 12 (with "Vespa" s/n 6, these being flown by Maj. Jorge Jordán Mercado, Cpt. José P. Coello and Tte. Luis Paravicini, respectively) were credited with destroying several key Paraguayan machine gun positions on 31 July 1932. Cpt. Rafael Pabón also claimed one of the earliest air-to-air victories of the war when he downed Paraguayan Potez 25 s/n 6 on 4 December 1932 while flying "Bolivian Scout" s/n 96. Another "Bolivian Scout," flown by Cpt. Ernst, was credited with the destruction of Paraguayan Fiat CR.20bis s/n 11-1 on 12 June 1933, as related earlier.

After the loss of two further aircraft in accidents, as mentioned (apparently s/n's 54 and 59) only s/n 8 and 96 survived into 1935, when s/n 96 also fell victim to an accident - possibly attributed to materiel failure. The last "Bolivian Scout," s/n 8, was still on strength with Grupo II at Villa Montes at the time of the Armistice in 1935, although it has also been reported to have been written-off enroute to Currurenda in November 1934 when, during a forced landing in the wild, its undercarriage was sheared off completely. This has led to speculation that this s/n was indeed used twice, the second time on one of the "spares" Type 143s.

Bolivian Vickers Type 143s were doped aluminum overall, with the exception of the undercarriage struts and interplane and cabane struts, which were apparently painted black. Standard Bolivian national insignia was carried in four wing positions and on either side of the rear fuselage, and the rudder stripes were standard as well. However, at least two of the aircraft, all of which carried individual serials forward of the fuselage national insignia in fairly large block numerals, also had a stylized Condor superimposed over the individual serials. These included s/n 9 and possibly s/n 8.

The "Bolivian Scouts" rendered good service under very rugged conditions and, while not as glamorous as the stylish "Hawks," were highly regarded even late into their six year Bolivian service careers.

Never before illustrated, the "Condor Boliviano" adorned the fuselage side of at least two Bolivian Vickers Type 143 and, in this case, is shown about to descend on "Leon Guarani" (the Paraguayan Lion,) the face of which appears to be the Paraguayan General José Félix Estigarribia! (Ramiro Molina Alanes).

WIBAULT TYPE 73 C.1

On 20 December 1927, at the express recommendation of the French Mission in Paraguay, the Government signed a contract with Charles Leonard Palabón House (Ateliers des Mureaux) in France for the purchase of seven Wibault Type 73 C.1 monoplace de chasse aircraft at a cost of 1,815,272 French Francs.

Following a rather lengthy delay (which has been attributed to labor and nationalization problems then on-going in France) the aircraft finally arrived in crates in Asunción in July 1929. Amongst the sales features of the rather diminutive fighter were its all-metal construction, ease of maintenance and ability to be erected or disassembled rapidly. Besides Paraguay, Brazil and Poland ordered essentially identical examples at about this same juncture.

Once erected at Campo Grande, these received serials "1" to "7" and were immediately pressed into service to train Paraguays first pursuit aviation cadre, a total of 13 of whom were trained on these aircraft between 1929 and 1931. The aircraft eventually became the mounts for the 1a Escuadrilla de Caza, although the exact date of constitution of this unit is not known.

The first 18 months of Wibault service in Paraguay, despite high utilization, were remarkably untroubled - in marked contrast to what was to follow. However, the first aircraft lost (s/n 4) was in 1931 near Asunción, when Tte. Ernesto Quiroz, while flying at very low altitude (possibly stunting for the benefit of family or a girl friend) struck the roof of a house with his undercarriage, crashing immediately in the suburb of Ypacaraí with the loss of his life and the aircraft.

On 24 August 1932, Paraguayan Fuerzas Aéreas Nacionales personnel at Campo Grande were alerted to the fact that an unknown aircraft had intruded into Paraguayan airspace and was approaching that main Paraguayan aerodrome. As hostilities had broken out full scale with Bolivia in June of that same year, it was immediately thought that this must be a far-ranging Bolivian re-

Paraguay had high hopes for the state-of-the-art Wibault Type 73 C.1 fighter but, like its other choice, the Fiat CR.20bis, this type was also beset by ceaseless powerplant problems. Here, Lt. Soldan is shown in front of an un-serialed Wibault at Ñu-Guazu in 1932. (Prof. Carlos Pusineri Collection).

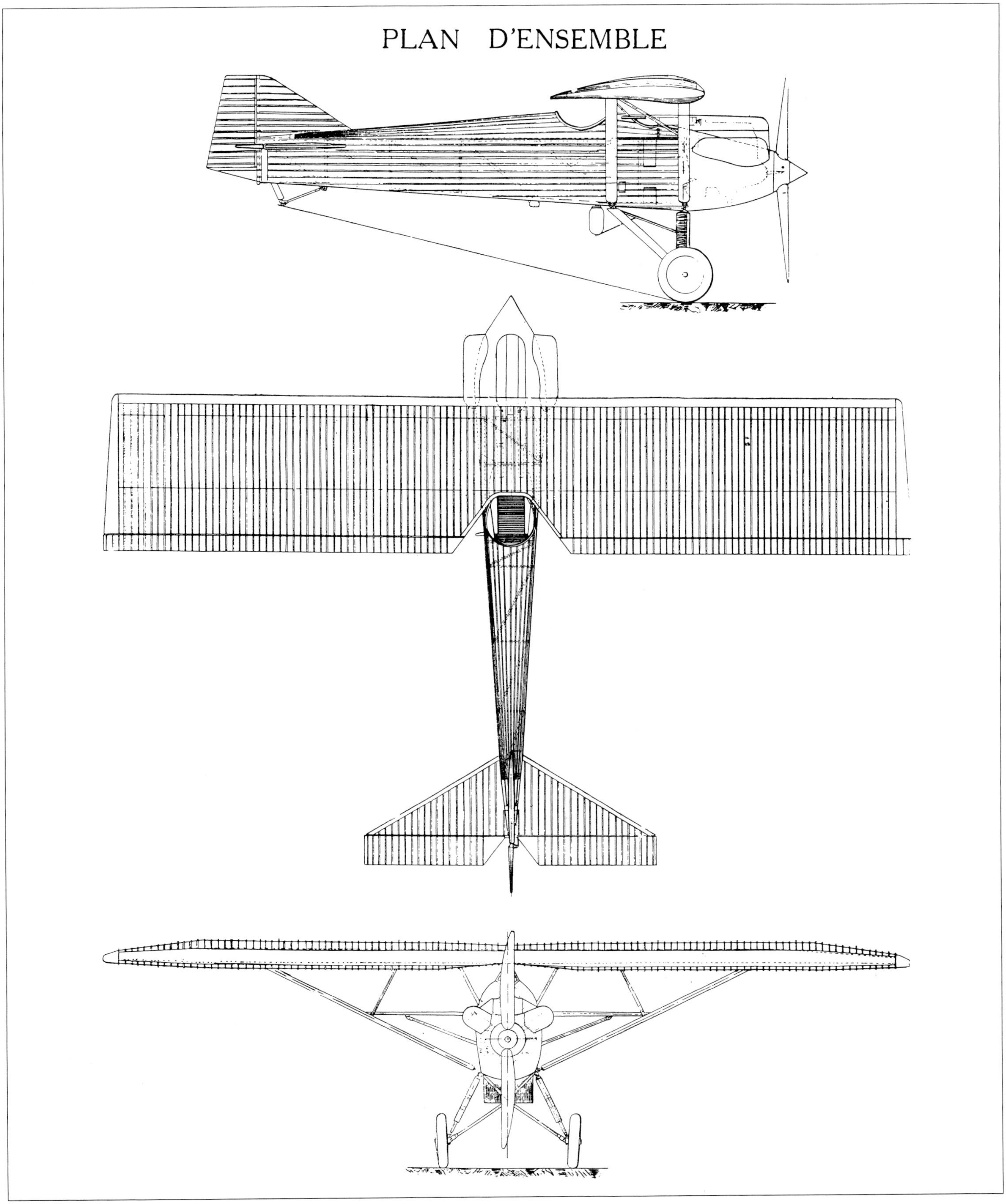

Three-view drawing of the Wibault Type 73 C.1 fighter (Wibault).

Lineup of virtually every airworthy Paraguayan service aircraft at Ñu-Guazu aerodrome in 1932, right before the outbreak of total hostilities. At extreme left is the exotic Savoia S.52, then all of the Wibault Type 73 C.1s. These were probably painted in the overall French green camouflage color of that period.(Lt. Gonzalo Palau Collection).

connaissance or bombing aircraft. Immediately, Wibault 73 C.1 s/n 5, piloted by Cpt. José Atilio Migone and s/n 6, flown by Tte.2° Walter Gwynn, scrambled to intercept the "bandit." The Wibaults pursued the unidentified aircraft for nearly 200 km before forcing it to land near the city of Belén. It was then discovered that the "bandit" was none other than a Chilean-assembled Curtiss D.12 "Falcon" piloted by an Argentine civilian pilot, Guillermo Hillcoat. This aircraft was amongst a batch of nine such machines that had been sold to a rebel group in Brazil then engaged in a bitter civil war. The aircraft had landed unscathed and had been immediately seized by Paraguayan infantry on the spot, was flown back to Asunción, and incorporated into the transport unit of the Arma Aerea. While an exciting event, this entire incident appears to be staged, however, as related in the section of this account devoted to the Curtiss D-12 "Falcon."

A total of six Wibault Type 73 C-1s were airworthy, therefore, when open hostilities commenced. Two of these, s/n 1 and 2, were almost immediately deployed to Isla Taguató Base, in company with Potez 25 A-2s s/n 4, 5 and 6 in August 1932. As soon as they had settled in there, the Wibaults started flying reconnaissance and escort missions with the Potez 25s over the Bolivian positions.

Just before the battle for Boquerón, all of these aircraft were redeployed to the Isla Poí Base, where they joined two other Potez 25s and a third Wibault (s/n 3). During the battle (usually cited as having taken place between 9 and 29 September 1932) Wibault s/n 1, 2 and 3 flew a total of 27 armed reconnaissance missions either solo or as escort to various numbers of Potez 25s.

The Paraguayan Wibault Type 73 C-1s all had the 450hp Lorraine-Dietrich 12Eb 12-cylinder, water-cooled engines, which proved to be very troublesome during the very intense utilization during the first months of the war. A report written during the war by Tte. Agustín Pasmor, Maintenance Officer for the entire Arma Aerea, stated bluntly that the Wibaults radiators (which were suspended beneath the cockpit, aft of the undercarriage, on the bottom of the fuselage, presenting obvious field maintenance problems) simply weren't large enough to cool the engine in the intense heat experienced in the Chaco region. When the Lorraine-Dietrich engine was functioning at its highest RPM (which was often the case in combat situations), the radiator coolant would start to boil, causing the opening of the radiator cap and the resultant loss of vital coolant. When this occurred, the Paraguayan pilots were instructed to land immediately and this was exactly what happened to s/n 1 and 2 while escorting Potez 25 s/n 1 on a reconnaissance mission over the Boquerón area on 30 September 1932. S/n 1, flown by Tte.1° Gregorio Morínigo had to make a forced landing near the Fortín Pozo Valencia where it was damaged beyond repair, although Morínigo survived. S/n 2 had exactly the same problem but landed safely near Boquerón itself. This aircraft could have been saved, but, following his landing, Tte.2° Luís Escario couldn't get help in moving the aircraft, which was overtaken by a fire which had started in a nearby pasture, which consumed the aircraft.

In spite of these serious engine problems, Wibaults s/n 5 and 7 were deployed to Isla Poí in October 1932, as the Paraguayans had little choice but to use what they had. S/n 6 remained at Campo Grande in Asunción undergoing repairs. Thus s/n 3, 5 and 7 continued flying escort missions with the redoubtable Potez 25s over

Wibault Type 73 C.1 serial number 5 in the field at Isla Poí aerodrome in August 1932 after the commencement of hostilities, Lt. Gregorio Morinigo posing before his mount. (Lt. Gonzalo Palau Collection).

Another view of a trio of Wibault Type 73 C.1s at Isla Poí in September 1932. Inexplicably, s/n 5 is followed in line by one serialed 23, while s/n 6 in the rear is missing its prop spinner. (Museo de Historia Militar via A. L. Sapienza).

Saavedra, Muñoz, Toledo, Arce, Corrales, Yujra and Cabo Castillo, in addition to many un-named places, between October 1932 and March 1933, fortunately without any more engine-related incidents, the crews having apparently been counseled to use maximum power only in extreme circumstances. During the raids over Saavedra and Muñoz escorting Potez 25s, Wibaults s/n 5 and 7 executed air-to-ground attacks against Bolivian Vickers "Vespa III" and "Bolivian Scout" aircraft, as well as several Junkers F 13 transports which were deployed there. The Bolivian Type 143s could not take off due to the extent of the damage to them which was apparently caused by the 7.7mm Vickers guns of the Wibaults.

In March 1933, the surviving Wibault 73 C-1s were all flown back to Asunción to be replaced, to the relief of the pilots, by the newly acquired Fiat CR.20bis fighters. During the remainder of 1933, Wibaults s/n 3, 5, 6 and 7 were used as fighter-trainers and for local air defense. Due to the commonality of the Lorraine-Dietrich engines used in both the Wibaults and the Potez 25s, the Wibaults were finally grounded entirely in order to provide their engines for the far more urgently needed Potez 25s. It is known that at least one of the Wibaults, s/n 7, remained in airworthy condition as late as the beginning of 1935, but they never returned to the active war theater again.

Paraguayan Wibault fighters were all painted a very dark green (approximating Humbrol HU116/FSN34079, including the engine cowling) with the Paraguayan rudder markings as usual and Paraguayan roundels in all four wing positions. The individual serial numbers, in white stylized numerals, was carried on both the fuselage, just aft of the cockpit, and underside of the wing inboard of the national insignia.

Not a single Wibault remained airworthy to participate in the Victory Parade that took place on 20 August 1935.

For the record, the first 13 pursuit pilots trained on Wibault Type 73 C-1s were as follows:

Maj. José Atilio Migone
Cpt. Leandfro Aponte
Cpt. José Gregorio Morínigo
Cpt. Bernardino Caballero Alvarez (later k.i.a. in a CR.20bis)
Cpt. Román García
Cpt. Walter Gwynn (later k.i.a. in a CR.20bis)
Cpt. Tomás A. Rafinelli
Cpt. Luís Escario
Cpt. (Navio) Victor Urbieta Rojas
Cpt. (HC) Vladimir Porfenenko (a Russian volunteer)
Tte.2° Orlando Salerno Netto
Tte.2° Abelardo Bertoni
Tte.2° Abdón Alvarez Albert

According to at least two sources, Bolivia had also entertained the notion of acquiring a variant of the Wibault parasol monoplane fighter at one point, apparently at the urging of its French Mission. These are reported to have been Type 72 C1s, which differed only in engine type from the Paraguayan aircraft. However, this deal apparently fell victim to the better terms and performance guarantees offered by the Vickers consortium.

Another view of Wibault Type 73 C.1 at Isla Poí in 1932 showing the position of the individual aircraft serial number under the port wing and use of conventional Paraguayan national insignia roundels (Ministerio de Defensa Nacional).

SAVOIA S.52

Paraguay's first pure fighter aircraft, the solitary Savoia S.52 acquired sometime in 1927, arrived in the country as a demonstrator, and was almost immediately acquired by the Government in order to give the Escuela de Aviación Militar students some fighter "experience" prior to the anticipated arrival of the Wibault 73s.

Originally developed in Italy in 1924, the S.52 (sometimes also cited erroneously as the SM.52) was one of the earliest all-metal structure biplane fighters, although it was of exceptionally light construction. Savoia, known for its line of (primarily) flying boats during the 1920's, seems an unlikely design bureau to have produced such a fighter.

Paraguay's aircraft was apparently utilized as a fighter-trainer at Campo Grande (also known by its native Indian name, Ñu-Guazú) from delivery through at least 1929 and, from then on, on an intermittent basis due to periods of unserviceability (one of the problems inherent in operating aircraft of a solitary, unusual design) its 300hp Hispano-Suiza H.S.42 engine being unique in the service inventory. Despite its frail appearance, the S.52 was actually quite sturdy, and was noted for its aerobatic performance. Its best speed was 260km/hr and it could climb to 5,000m in 15 minutes, 30 seconds.

Nominally capable of being armed with two 7.7mm machine guns, the S.52 is not known to have ever strayed far from Campo Grande and, if it ever carried a serial number, this has been lost to history. The aircraft remained on strength, however, until 8 May 1933, when it suffered an irreparable accident, probably at Campo Grande. According to most accounts, the aircraft was flown almost entirely by Tte. Emilio Rocholl.

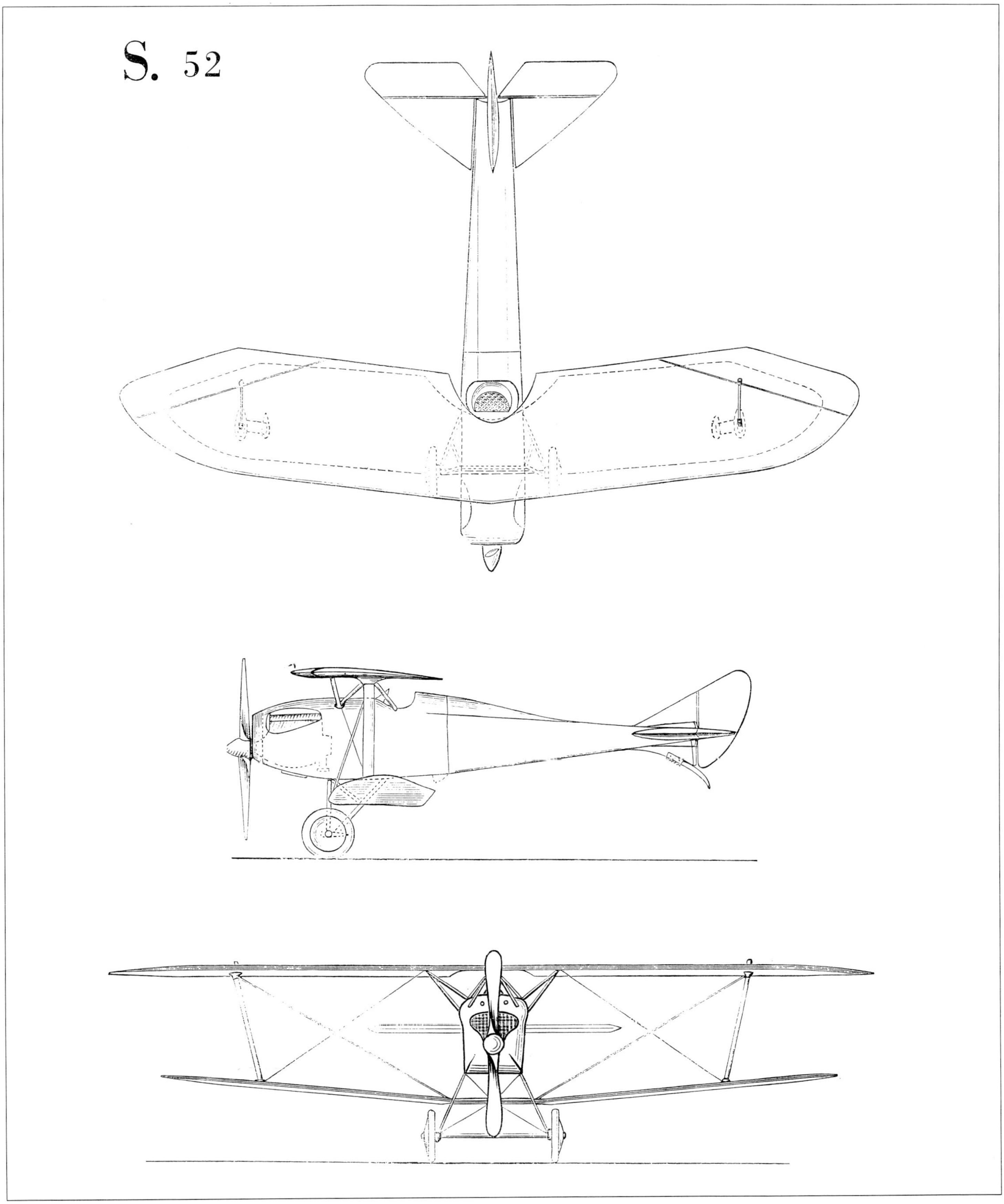

Three-view drawing of the Savoia S.52 fighter (Savoia).

Little is known of the diminutive Savoia S.52 fighter acquired by Paraguay in 1927, except that it served primarily as a fighter-trainer and appears to have been flown by relatively few Paraguayan service pilots. One of these was Lt. Emilio Rocholl, pictured here. It suffered irreparable damage at Asunción on 3 May 1933 and was not flown again. So far as is known, it never carried a serial number of any kind. (Antonio Sapienza Collection).

Available photos of the diminutive S.52 show that the aircraft was apparently painted some dark color (possibly the same shade green as worn by the Wibault Type 73 C1s) and it had standard Paraguayan rudder markings. No fuselage roundel or serial number is known, however, and details of wing insignia remain unclear.

The end of the solitary Savoia S.52, which crashed at Ñu-Guazu on 8 May 1933 while piloted by Tte. Emilio Rocholl, who received minor injuries in the incident. Apparently no effort was made to repair the aircraft, which appears to have suffered comparatively little damage (Tte. Emilio Rocholl).

GOURDOU-LESEURRE LGL.32 C-1

There are several reports that Bolivia acquired at least one Gourdou-Leseurre (sometime given as Loire-Gourdou-Leseurre, hence the designation LGL.32 C-1) parasol fighter monoplane sometime around 1926.

Similar in appearance in many ways to Paraguay's Wibault Type 73 C.1s, the LGL.32 C-1 was one of the winners of a 1925 French Air Force fighter competition. The type was the first to be produced by the merged Gourdou-Leseurre and Ateliers Chantiers de la Loire firms which had merged in 1925. Some 350 of these aircraft were acquired by the French, with whom it went into service circa 1926-27, remaining in front-line service until 1930. Examples were also exported to Romania (at least 50), Turkey (12) and Spain (8).

Although not mentioned in official Bolivian histories, at least two French annuals mention the type in Bolivia, as does JANE'S. This would have made the LGL.32 C-1 Bolivia's first genuine single-seat fighter type, but the suspected solitary example was apparently wrecked or lost otherwise some time before active hostilities with Paraguay actually commenced, and thus has only a "curiosities" place in this narrative.

4

Other Types-Transports, Ambulances, Trainers

THE TRANSPORTS

JUNKERS F 13L AND F 13W

Bolivia enjoyed the immeasurable advantage of having a very capable and experienced air transport capability available to its military machine from the very earliest stages of the conflict. Here, virtually the entire fleet of Lloyd Aeréo Boliviano (known as LAB) is lined up on 18 May 1928 after mobilization, consisting of six Junkers F 13s. Note the Bolivian tricolors on the rudders, the sole concession to a quasi-military identity (LAB via Juan Arraez Cerdá).

Flying boats and float planes were used extensively by Paraguay, but only a few Bolivian aircraft were water-borne. This LAB Junkers F 13W, "Beni" was one such example, shown here flying over the trackless Chaco in 1928. (LAB via Juan Arraez Cerdá).

Bolivia's war effort was aided significantly by the presence in the country of the Lloyd Aéreo Boliviano, the entire fleet of which in existence as of 7 March 1928 was incorporated into the Bolivian Army via an executive decree of that date.

At that time, the LAB fleet consisted of six Junkers F 13s, of which all six started out as F 13L land-planes, although two were apparently converted to F 13W float-equipped aircraft sometime after April 1928. Mounted with a mix of 300hp BMW IV and Junkers L-5 engines, these aircraft operated throughout the Republic and into the war zone on behalf of the Bolivian war effort, although several were lost to accidents and crashes in the period between the date of the Executive Decree and the armistice.

The F 13s operated with standard Bolivian rudder stripes but did not carry Bolivian military roundels on the wings. Instead, they had

Another of the LAB F 13Ls, "Mamore" was converted to float configuration as an F 13W and is shown operating in the Chaco with "Beni" at mid-stream. Just visible on the wings of "Beni" are the bands containing the Bolivian national colors, used to aid in identifying the aircraft to friend and foe. (Hagedorn Collection).

This view of two LAB Junkers F 13Ls, believed taken shortly after mobilization of the LAB fleet, shows again the early placement of the Bolivian colors on a white band under the outboard wing of the middle aircraft, believed "Mamore." (Courtesy Georg von Roitberg).

large Bolivian flags painted on white panels on either outer upper and lower wing on most of the aircraft, although some apparently flew with only the rudder stripes and the individual aircraft name on the fuselage. Others had the flag on the white panel on the lower side of the wing only. Contrary to popular belief, these aircraft did not receive these names until after they were drafted into military service, the Christening ceremony having taken place on 18 May 1928. They were:

Variant	Name	Notes
F 13L	"Charcas"	Was wrecked by a storm in 1934. Had 300hp BMW IV.
F 13L	"Illimani"	Still in service as of 23 May 1933. Had 300hp Junkers L-5.
F 13L	"Oriente"	Often given in error as "Orient." Crashed 27 August 1929.
F 13L	"Oriente II"	Crashed in November 1929.
F 13L/W	"Mamoré"	Had 300hp Junkers L-5. Still in service 23 May 1933.
F 13L/W	"Beni"	Had 300hp BMW IV. Still in use at 23 May 1933.

A period LAB advertisement showing the airlines route network as of the time of the Chaco War as well as in-sets of some of the lines F 13s involved in the war effort. (LAB)

A hitherto unknown, final Junkers F 13 was acquired by LAB for war service in July 1929. Named "Edelfalke," a decidedly Germanic name compared to the earlier fleet names, this aircraft was used to perform reconnaissance flights for no less than the Chief of Staff into the Chaco. (Junkers).

It appears that possibly three of these aircraft were c/n's 534, 634 and 788, as two of these survived the war to return to LAB service, and were still on the fleet list as of January 1942.

However, one and possibly two other F 13Ls served with the Bolivian transport element during the Chaco War. One of these was named "Chaco" (which crashed some time in 1928, although this aircraft may have been one of the earlier machines re-named for propaganda purposes) while another, not received in-country until July 1929, carried a purely German name (at least at first), "Edelfalke."

Besides innumerable transport flights from La Paz to the front and other points in the Republic during the war, at least two of the aircraft actually performed combat support and reconnaissance duties. The first of these was "Illimani," which provided supplies and ordnance support for the Breguet 19 "Potosi" which attacked Paraguayan positions from Puerto Suarez in January 1929. "Edelfalke" also served, as a reconnaissance aircraft, carrying the Bolivian Chief of Staff into the Boreal region in July 1929 for a first-hand look at things.

It appears likely that the LAB F 13s actually performed at least some revenue producing commercial flights during the period between being drafted into National service and the armistice. It has often been implied that the fleet was given over to purely military support entirely for the duration, but this does not appear to have been the case, although there were certainly periods during which they were totally preoccupied with military support activities.

Although arguably acquired for use by LAB, the Junkers W 34ci "Vanguardia" shown here at the left of a lineup of military Breguet 19 A2s amd Caudron C 97s at El Alto, was in fact at this point for all intents and purposes a military aircraft. (Ramiro Molina Alanes).

JUNKERS W 34ci

Besides the venerable and faithful Junkers F 13s drafted into government service from LAB, at least three Junkers W 34ci transports eventually also saw Chaco War service with Bolivia, and rendered excellent services almost for the duration.

So far as can be determined, the first W 33/W 34 types acquired for Bolivia was the aircraft referred to as the "Junkers `Bremen' type" purchased via popular subscription by 23 December 1929, which was in Germany as of that date awaiting delivery to a Bolivian crew (CPT Lucio Luizaga and CPT Horacio Vasquez) who proposed to fly it from Berlin to La Paz. However, in view of the worsening situation with Paraguay, the General Staff instead directed that the flight be canceled and that the aircraft be shipped home to Bolivia forthwith for use in "the ongoing training program." The aircraft was apparently shipped, via Hamburg, around 20 January 1930, which coincides with other supporting data. Oddly, the fate of the two pilots is not known, as neither name shows up in any published listing of Bolivian war veterans as aircrew. Although it cannot be conclusively proven, this aircraft seems most likely to have become the one named "Vanguardia" and which apparently later gained the military serial 101.

All three of the W 34s acquired by Bolivia had 525hp Pratt & Whitney "Hornet A" engines, but differed in other minor respects, suggesting that they were acquired piecemeal, which seems likely. The other two aircraft seem to have followed, shortly after "Vanguardia," later in 1930, and the three seem almost certainly to have been c/n 2607, 2608 and probably 2609.

The second pair of W 34s were named "Tunari" and "Presidente Siles" (which apparently also carried the serial "76" at least briefly) but eventually these two aircraft seem to have been re-serialed as 102 and 103 in the military series.

By 23 May 1933, all three were still very much active and engaged in not only intensive transport activities, but limited light bombing as well. T. W. Smith of Gran Chaco Adventure specifically cites s/n 101, 102 and 103 as having been fitted with Busche

Similar in many respects to the somewhat more refined Junkers K 43 bombers, the W 34ci "Vanguardia" was apparently used not only as a dedicated transport, but as a bomber as well. It is seen here at Villa Montes in 1934, well into its wartime service (LAB via Juan Arraez Cerdá).

bomb sights at Villa Montes some time in late 1934 and, together with the Junkers K 43s, these formed the equipment of the famous Punta de Alas bombing unit which operated so effectively from that base for the remainder of the war.

Unlike the K 43s, however, the W 34s are not known to have had any gun armament, the bomb sights and underwing racks having been their sole militarization.

Number 103 (the former "76," apparently still named "Presidente Siles") was wrecked in spectacular fashion at Santa Cruz on 25 May 1935 in a non-combat related accident, and so did not survive the war. Both "Vanguardia" and "Tunari" survived the war, however, to return to LAB service as civil transports, the first still in regular service as late as October 1938 and the latter until at least January 1942, after which it is believed to have gained the early Bolivian civil registration CB-23.

Another view of Junkers W 34ci "Vanguardia" also in 1934, but by which time its engine mount area, previously painted black, had been cleaned up and all paint removed from metal surfaces. (Antonio Sapienza Collection).

Perhaps the most prized Bolivian assets of all were the LAB Junkers Ju 52/3m transports, one of which is shown here being prepared for a crude-field take-off at Machareti, north of Villa Montes, in 1933. Note the absence of speed rings on the outboard engines. This is probably "Chorolque." (LAB via Juan Arraez Cerdá).

JUNKERS Ju 52/3mde and Ju 52/3mge

While their single-engined forbearers, the Junkers F 13s and W 34s, carried the torch for Bolivia between 1928 and 1932, the Bolivian cause and troop morale were boosted greatly in December 1932 by the arrival of three tri-motored Junkers Ju 52/3m transports (a fourth aircraft, c/n 4061 "Bolivar," sometimes cited as a Chaco War veteran, in fact did not arrive in-country until sometime in February 1935, and seems to have played only a minor part in the proceedings).

While these aircraft may arguably rate pride-of-place in this discussion of transport aircraft fielded by Bolivia, their comparatively late arrival on the scene and intentional husbanding as coveted capital items by Bolivia's leadership combined to make their contributions - although great - something less than they might have been otherwise.

One of the most controversial questions to arise from the Chaco War was, simply, why Bolivia did not find the means to convert these heavy-lift aircraft, capable of a range with load of some 568 miles (915km), into heavy bombers to support the hard-pressed Bolivian armies in the field - or, in view of the lengths that Bolivia later went to in attempting to acquire Curtiss-Wright BT-32 "Condor" bombers for the intent purpose of bombing Asunción - why the Ju 52s were not engaged in a similar attempt.

On the surface of it, the means and capability to convert the Ju 52s to bombers were probably available to Bolivia. However, the

Bolivian and German (LAB) officers pose with another of the veteran Junkers Ju 52/3ms, "Juan del Valle," at Villa Montes in 1935. Although operated intensively, the aircraft were well maintained and performed admirably. (LAB via Juan Arraez Cerdá).

Although some foreign pilots are known to have served with the Bolivian military out-right, German pilots who flew for LAB on its wartime transport services are less well known. Here, Herman Schroth (right) and Emil Fastner (left) pose at Villa Montes in 1935 as one of the Ju 52/3ms is serviced. Note the band under the out-board wing containing the Bolivian colors. (LAB via Juan Arraez Cerdá).

Ju 52/3m "Huanuni", for reasons unknown, received full camouflage "warpaint" during its wartime service. It is seen here at Muñoz in 1933 delivering priority supplies. (LAB via Juan Arraez Cerdá).

"Huanuni" carried virtually everything needed, anywhere in the war zone. Here, the gun carriage of a light field piece is hand-loaded aboard the aircraft. (Hagedorn Collection).

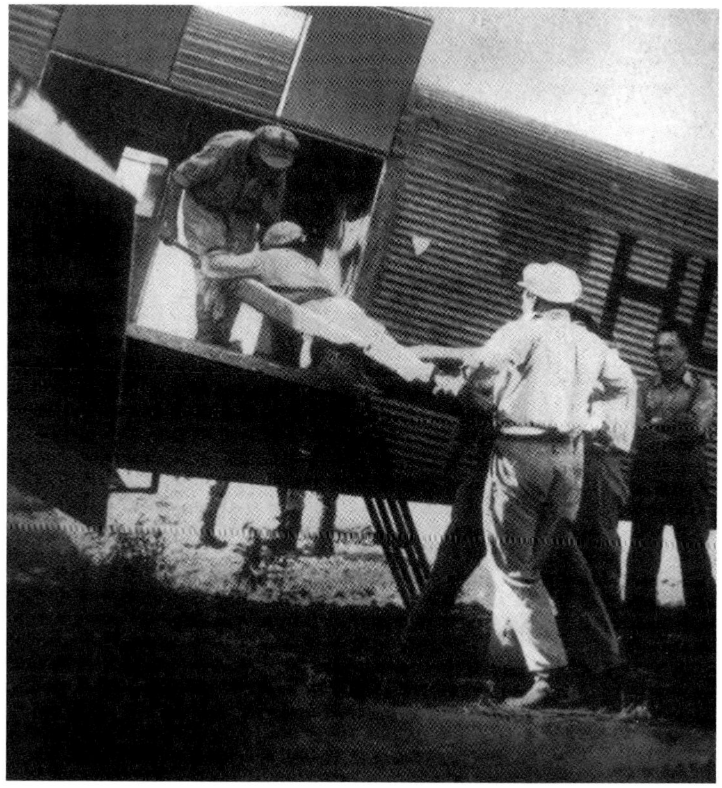

Perhaps its most vital role during the war was casualty evacuation - one of the very few morale boosters for the hard-pressed Bolivian soldiers in the trenches. This, again, is the camouflaged Ju 52/3m ostensibly of LAB, named "Huanuni." (Hagedorn Collection).

answers to these puzzling questions seem to actually point in a rather different direction. At least two (and possibly all four) of the Ju 52/3ms that Bolivia eventually operated during the war were acquired via the largesse of Sr. Simon I. Patino, the tin magnate. The actual conditions attached to these aircraft in this process are not known (and may never be known), but it seems clear that Sr. Patino and LAB both wanted to protect this very considerable investment for post-war use and, while they were willing to see the aircraft engaged in heavy transport, casualty evacuation and similar actions, even the official Bolivian Air Force history of the war stresses that ...."the aircraft were great assets and, consequently, were granted the greatest of care in dispatch to the front, and then only into areas where enemy air activity was assuredly infrequent." Finally, while it may have been just possible for the Ju 52s to have reached Asunción from the very farthest point forward of the Bolivian advance, this would have required a prodigious supply effort to deliver the required fuel and ordnance, and even then, the aircraft would have had to operate from extremely crude, short fields at

Full-length view of the unusual, and fairly effective camouflage scheme as applied to "Huanuni" during the war, as it once again is seen unloading supplies onto vintage trucks. Note the three unattended large caliber aerial bombs on the ground to the rear of the aircraft. It has been suggested that this Ju 52/3m was used at least once as a heavy bomber, but no evidence has surfaced to confirm this allegation (Hagedorn Collection).

An unidentified LAB Junkers Ju 52/3m (but probably "Chorolque") showing the identification bands under (and also over) the outer wing bearing the Bolivian colors, but also on the upper rudder, the only known instance when this occurred on a Ju 52 during the war. Note the fuselage of a derelict F 13 in the left background (Hagedorn Collection).

It is not clear which of the Ju 52/3ms this is, but it bears evidence of some sort of upper surface camouflage attempt. Note the "speed ring" color differences on the upper and lower sides. This may be "Huanuni" at a different stage of the conflict (Courtesy Georg von Roitberg).

Yet another variation on the LAB Junkers Ju 52/3m theme during the Chaco War, this aircraft (probably "Juan del Valle") appears relatively clean bare-metal, aside from the black forward fuselage and underside - but has no speed rings at all! The outer wing identification band is evident, however (Ramiro Molina Alanes).

almost their maximum range - without escort. In short, it was too risky.

The Ju 52s acquired by Bolivia were c/n's 4008, 4009, and 4018 (and, as stated earlier, c/n 4061, which arrived late in the conflict, and which seems to have been fitted out in a VIP configuration). The first two had 575hp Pratt & Whitney "Hornet D1" engines, while the third had 525hp "Hornet A1s" a key recognition point in distinguishing these three high-time veterans apart. At least two of these bore dedicated camouflage during the conflict (a relative novelty during the war) as well as their individual aircraft names "Juan del Valle," "Huanuni," and "Chorolque." So far as can be determined, none bore any military serials, and the only concession to national markings was the Bolivian tri-colors on the rudders, in varying dimensions, on all aircraft.

"Bolivar" which, as mentioned, appears to have had luxury appointments, was used to deliver members of the Neutral Commission to the Armistice talks in May (when it was described as the "new" Ju 52) and again on 16 September 1935 while no less than the President of Bolivia himself is known to have traveled to Villa Montes at least once on one of the Ju 52s in late November 1934.

All three of the early Ju 52s operated without wheel "spats" during the war, and have been seen with and without Townend rings on their P & W engines. At least one (probably "Huanuni," which seems to have been hazarded the most often at the front) also had Bolivian colors painted under the outer wing panels, similar to the style known on several of the F 13s. Known to have been delivered natural metal, "Huanuni" is definitely one of the aircraft which was later camouflaged.

All four aircraft survived the war and passed almost immediately to LAB.

TRAVEL AIR MODEL S-6000-B

Paraguay also had air transport and ambulance needs, but did not enjoy the luxury of an established, well-equipped internal airline to nationalize. Instead, agents acquired used but adequate transports where they could. This Travel Air Model 6000, serialed T-9 and named "Nanawa," still bears its former U.S. airline name "Boston," besides full Paraguayan ambulance and national markings in this 1933 photo at Ñu-Guazu. (Archivo Cnel. Av. (R.S). A. Pasmor).

Paraguay never enjoyed a valuable transport fleet such as was available to Bolivia but, instead, made do with a truly amazing assortment of single examples (with but one exception) of mainly second-hand and multi-purpose aircraft, only one of which was multi-engined.

Amongst these, it may be fairly claimed that a pair of single-engined, monoplane cabin Travel Air Model S-6000-Bs were the most important.

Often thought to have been acquired in neighboring Argentina, the two S-6000-Bs were actually former Inter-Cities Airlines, Inc. and Transcontinental Air Transport (TAT) aircraft, apparently exported quite illegally to Paraguay (in one case) in mid-1933 when the two airlines divested themselves of some equipment due to the on-going depression.

The most popular version of a family of buxom Travel Air monoplanes, the 6000-B was originally approved by the U.S. Department of Commerce under Approved Type Certificate No.130 in March 1929 (modified in July 1929 to read S-6000-B, which meant that the aircraft was also approved to be mounted on twin-floats as a seaplane) with the reliable 300hp Wright J6 engine. Capable of quick-change interiors seating up to six or, stripped, used as a freight carrier, the S-6000-B had a very respectable performance and pleasant handling characteristics.

Paraguay acquired its two examples primarily for use as ambulances, through the auspices of the Junta de Aprovisionamiento Nacional, which had traveled to the U.S. for the express purpose of acquiring ambulances at the end of 1932, and both had been fitted with upgraded 330hp Wright engines. The two aircraft were c/n 6B-2011, formerly NC624K and probably c/n 1029, NC9815 (which carried the name "Boston" with T.A.T., this aircraft having been exported ostensibly to Argentina on Export Certificate of Airworthiness E-5190). Of the pair, c/n 6B-2011 became Paraguayan serial T-2. Interestingly, the U.S. registration was canceled, officially, as "expired, not renewed" on this airplane effective 1 November 1933, with no other disposition given in its record. The second aircraft was given Paraguayan serial T-9 and was named "Nanawa."

As an ambulance, T-9 "Nanawa" (which at one point retained, ironically, its former airline fleet name "Boston" on the nose at the same time) was painted black overall but, although it carried standard Paraguayan rudder markings, instead of roundels, had white discs in all four wing positions and on either side of the rear fuselage with red crosses on them. The serial, T-9, was carried in white characters high on either side of the rear fuselage. T-2 on the other hand was painted aluminum dope overall (except for those portions of the forward fuselage which were metal) and had standard Paraguayan national roundels in all four wing positions and standard rudder marks. The serial was in black characters on the mid-fuselage either side.

Interestingly, it is also alleged that a Travel Air monoplane was flown into Paraguay on 17 May 1933 from San Fernando, Argentina, by one Sr. Honer and a Mr. Hillcoat as pilot. It is alleged to

Another view of Travel Air Model 6000 T-9 named "Nanawa" showing additional markings details and two undoubtedly grateful casualties in the foreground about to be air-evacuated. (Ministerio de Defensa).

A second Travel Air Model S-6000B was also acquired by Paraguay and coded T-2 and, by the time of this April 1939 photo, had also gained the name "Nanawa." Note the doped aluminum colored Potez 25 second in line with the two cabin slit windows and enclosure for passenger carrying duties. Next in line is a Fleet Model 2 then three Breda Ba 25 trainers (Archivo Cnel. Av. (R.S). A. Pasmor).

have been purchased for Paraguay in Argentina by a Sr. Espinosa, but nothing more is known about this aircraft. As both known aircraft arrived via boat at Buenos Aires, Argentina in mid-1933, this aircraft may in fact have been one of the two described above, and an unknown Argentine "middle-man" may have been involved.

Both known aircraft survived the war (although one appears to have been disassembled for a time), after very hard use, and were still on strength with the service as late as August 1944. By 13 September 1945, they appear to have passed to the military operated airline LATN, with whom both remained in use as late as 31 October 1947.

Another view of Travel Air S-6000B T-2 "Nanawa" at Ñu-Guazu in 1939. The Travel Air's were amongst the last of the Paraguayan Chaco War veterans to remain in active service, both aircraft still being on strength as late as October 1947. (Archivo Cnel. Av. (R.S). A. Pasmor).

Travel Air S-6000B T-2 at the forward airfield of Isla Poí in 1933, showing evidence of yet another instance of the unauthorized "star" type national insignia under the starboard wing. (Museo de Historia Militar via A. L. Sapienza).

BREDA 44

Sometimes given as the Breda Ba 44, this light, twin-engined transport aircraft bore more than a little resemblance to the British de Havilland "Dragon" of similar vintage.

Indeed, the Breda aircraft was powered by two 190hp Walter "Major-Six" engines (similar to the units used on the "Dragon") and could boast a respectable top speed of not less than 140mph with up to six passengers.

Acquired in March 1933, apparently a new-build airplane, the Breda 44 was also intended for service as a hospital aircraft, making its first flight as such into the zone of operations on 22 March 1933 serialed T-1.

Another exotic transport aircraft to see solitary though extensive service for Paraguay during the war was a Breda Ba 44, a de Havilland D.H.84 "Dragon" look-alike. Coded T-1 and named "Asunción," it is shown here with (at left) Lt. Enrique Dentice and Lt. Agustín Pasmor at Ñu-Guazu in 1933. (Archivo Cnel. Av. (R.S). A. Pasmor).

By 1934, Paraguay's Breda Ba 44 had undergone some markings changes, chief amongst which was a dark (probably black) fuselage underside, probably to help mask damage to the fuselage from operations in the rough. This photo was, indeed, taken at Isla Poí in 1934, a dangerous place for an unarmed transport! (Archivo Cnel. Av. (R.S). A. Pasmor).

Painted white overall (except the underside of the fuselage, which was apparently painted black) the aircraft also carried red crosses on white discs on the wings and standard Paraguayan rudder marks. T-15 had the honor of flying, in company with a Paraguayan Potez 25, to Villa Montes with the Paraguayan armistice delegation on 14 June 1935.

T-15 remained in service through WW2, passing to LATN by 31 October 1947, by which time it was noted as "in overhaul."

Obviously repainted once again, Breda Ba 44 T-1 had the distinction of carrying the Paraguayan delegation to the armistice talks at Villa Montes in 1935, where it is shown here, by which time, for reasons unknown, it had been re-serialed T-15. (Ministerio de Defensa Nacional).

Another view of the Paraguayan Breda Ba 44 while coded T-15 and bearing ambulance markings. Capt. Emilio Nudelman (first on left) poses here with other Paraguayan officers at none other than Villa Montes, Bolivia right after the cease fire on 14 June 1935. (Sgto. Ramiro Molina Alanes).

SAVOIA S.59bis

Often thought to have been a combatant aircraft, Paraguay's single Savoia S.59bis, although possibly used on several unarmed reconnaissance missions, had in fact been built as a four-passenger airliner version of this versatile Italian flying boat, and is not known to have ever been armed.

This well-traveled aircraft was purchased originally in the last half of 1928 by an Italian officer, Colonel Ernesto Colombo, for his taxi airline being established at Puerto Nuevo in Buenos Aires, Argentina, named Taxi Aereo. The aircraft was registered in Argentina as R-ACTZ. Although highly popular for flights across the mouth of the Rio Plata to Uruguay and sight-seeing over Argentina's capital city, an accident by a low-flying Argentine civil aircraft led to the prohibition of low-level flights carrying passengers over the city and Taxi Aereos fortunes immediately took a turn for the worse.

Colonel Colombo, learning that Paraguay might be in the market for durable aircraft, departed the second week of April 1929 for Asunción and, following the 1,350km flight (only the third flight ever between the two cities), and subsequent demonstration of the aircraft to Paraguayan Naval authorities, the aircraft was purchased outright (together with a Cant 10ter owned by Taxi Aereo, then still in Buenos Aires). The Italian crews (including Colonel Colombo and one of the pilots, Antonio C. Costagliola) were asked to stay on as well as "instructors," and, after strenuous service as a trainer between 1929 and 1932, the S.59 went to war as a transport.

Besides flights carrying ammunition, wounded troops, VIP's (as many as five, including the pilot, on one occasion) and dispatches, (over mostly waterless dry reaches), including rare night flights, the S.59 was wrecked near the end of 1933 on a return flight from Bahia Negra deep in the Chaco. After Herculean effort, the remains were carried and rafted out and rebuilt, innovatively enough, with many local wood product substitutes for smashed original Italian parts, and was returned to service. At around this point, she was named "Chaco," having carried the serial R-1 for some time.

Incredibly, the S.59 survived in service with the Navy as late as 1 July 1943, (contrary to reports it was dismantled in 1935) although by 6 May 1944 it was reported to have been junked.

The aircraft is known to have retained its original "Savoia Marchetti" titles on either side of its bows. It was painted doped aluminum overall otherwise, except the bottom of the hull, which was black, with a snazzy thin waterline just above the edge of the planing bottom (also repeated on the wing floats). It had standard Paraguayan rudder marks and the serial was carried in black characters on the rear fuselage, either side. It is also known to have had a special insignia of some sort on (at least) the port bow and at least one photo seems to show an anchor under the left lower wing inboard of the float.

Although frequently cited as a combatant aircraft, the solitary Paraguayan Navy Savoia S.59bis was only used as a transport and unarmed reconnaissance aircraft. Here, looking quite smart in this April 1929 view, the aircraft gained the serial R-1. (via Dr. Roberto Gentilli).

FORD MODEL 5-AT-D "TRI-MOTOR"

Bolivia was quick to recognize the high value of the transport capability provided by the Junkers F 13s and W 34s of LAB, and further enhanced this asset by acquiring the later Junkers Ju 52/3m tri-motor transports.

LAB apparently wanted something sooner to augment the single engined Junkers, however, as arrangements had been concluded in mid-1932 for the purchase in the U.S., via the Central Bank, of a single factory-new Ford Model 5-AT-D, c/n 5-AT-113, which carried U.S. civil registration NC 9654.

Having made its first flight in March 1932, Ford advised the U.S. Department of Commerce that, by 21 September 1932 (and probably some time prior to that date) the aircraft had been "sold into South America."

Indeed, after an epic ferry flight via the U.S. Army Air Corps base at France Field in the Panama Canal Zone, the aircraft, flown by Donald Duke and a mechanic named Willy Kettman (alleged to be German, and who was probably a LAB employee) finally arrived at La Paz. It was apparently almost immediately stripped of its passenger accommodations and pressed into service as a freighter.

Given the name "Cruz del Sur" the aircraft is illustrated in the official Bolivian Air Force history of the Chaco air war as having carried Bolivian national insignia, including rudder stripes and wing roundels. However, the only known photo of the aircraft in Bolivia shows the aircraft sans markings of any sort, with the exception of the "Ford" logo on the vertical fin and the U.S. registration across the top of the fin and rudder. This aircraft was lost at Villa Montes on 26 October 1932 when, with a very heavy load of fuel and artillery shells for the front, it apparently lost an engine on take off and was destroyed.

Although nominally LAB property, the single Ford 5-AT-D acquired by Bolivia in September 1932 operated exclusively in support of the Army in the Chaco until lost after only 35 days! Alleged to have gained full Bolivian military markings, it was named "Cruz del Sur" and was the former NC 9654. (Ramiro Molina Alanes).

CURTISS "ROBIN" C

Another oddity amongst the diverse second-hand transport fleet assembled by Paraguay was a single example of the Curtiss "Robin" C, which was utilized in a variety of light transport and utility missions from acquisition in October 1933 until after the war.

Although frequently cited as having acquired as many as three "Robins," only one has been positively traced that actually arrived and entered service. Utilized by the ad hoc Escuadron de Transporte, along with the other odd types, the aircraft arrived via Argentina and may have had a civil registration in that country. It is believed that the aircraft which actually arrived was c/n 272, the former U.S. Identified Aircraft No.8399. This aircraft was sold by Leigh Wade of Tri-American Aviation to a Juan Cassinera in Buenos Aires circa 7 October 1933. Wade is known to have been engaged in arms sales and since the single "Robin" acquired by Paraguay did not make its appearance until about this same juncture, the odds seem to favor this aircraft as the one in question.

Coded T-7 (the first use of this serial) the aircraft was also used on occasion as an ambulance as well as a VIP transport, and survived in Paraguayan service until 1938.

Before leaving the Paraguayan "Robin" it should be noted that a U.S. State Department document dated 21 April 1936, reporting on the arms sales to Paraguay during the war, stated that Paraguay had acquired three "Robins" during the war. The source for this report was not cited, and has led to speculation that two other aircraft - perhaps also purchased via Wade - could not be spirited out of Argentina before the authorities stepped in.

BREDA 15S

A rather odd looking aircraft, a single Breda 15S touring aircraft (sometimes given as a "Ba 15S") was in Paraguay as early as 1929 and, in several photos showing line-ups of Paraguayan military aircraft, is often seen in civilian guise bearing its Italian civilian registration, I-AAUG.

The aircraft was, however, apparently eventually incorporated into the Fuerzas Aéreas Nacionales and received serial T-8 - the next serial available after the Curtiss "Robin," which suggests that its date of acceptance into actual service must have come around the same time.

The aircraft's actual utilization is poorly documented. A high-wing monoplane with a very prominent 115hp Colombo S.63 engine and exhaust system projecting up and over the elliptical wing, the Breda 15S could have only had very limited durability, as its cabin could accommodate only two persons and the aircraft was reputed to be rather fragile under other than ideal conditions.

The aircraft was lost in a 1933 accident.

Another of the strange aircraft collected by the Paraguayan Arma Aérea was this Breda Ba 15S touring aircraft, shown in a military lineup still bearing its Italian civil registration marks I-AAUG!. It eventually gained military marks and the serial T-8. It is flanked by the hard-worked Morane-Saulnier 139 (Archivo Cnel. Av. (R.S). A. Pasmor).

CONSOLIDATED MODEL 21-C ("PT-11")

Paraguay's array of unusual aircraft included the solitary Consolidated Model 21-C demonstrator (c/n 14), which had been purchased on the spot from the Consolidated front man, Leigh Wade, when he was demonstrating it in South America. The colors of the aircraft at this point are uncertain, but the Paraguayans gave it the transport serial T-11 and it was used as a high-speed courier and VIP aircraft. No less than the President of Paraguay is in the rear seat in this 1934 photo. (Antonio Sapienza).

This view of Paraguay's Consolidated 21-C, taken at Isla Poí just prior to returning to Asunción in 1933, reveals that a camouflage pattern had been applied to the fuselage (at least) and a colored recognition band (possibly red) was painted under the lower wing just inboard and overlapping the national insignia. (Museo de Historia Militar via A. L. Sapienza).

In this 1933 photo of the single Paraguayan Consolidated Model 21-C (frequently cited in error as a "PT-11") the fuselage of the aircraft appears to have been painted a dark color, although the cowl ring and underside are light. It is seen here at Isla Poí. (Archivo Cnel. Av. (R.S). A. Pasmor).

One of the mystery aircraft of the Chaco War, a single aircraft identified in all Paraguayan documents as a Consolidated "PT-11" arrived in Paraguay some time in 1933.

Photographic evidence clearly shows that the aircraft in question was, indeed, an aircraft with essentially the same characteristics as the Consolidated PT-11/BT-7 series as used by the U.S. Army Air Corps.

However, as this aircraft obviously did not originate from U.S. Army Air Corps sources, it was much more likely that the aircraft in question was the solitary example of the Consolidated Model 21-C (c/n 14) that is otherwise unaccounted for. This aircraft was the Consolidated demonstrator and it was extensively marketed in South America by none other than Leigh Wade commencing in December 1932. So far as can be determined from available records, the aircraft never returned to the U.S. Given the depression era economics facing Consolidated, and the fact that Wade was associated with at least two other aircraft sales to Paraguay, this is almost certainly the aircraft in question.

Given Paraguayan serial T-11 (the first use), this aircraft actually served as a high-speed courier and communications aircraft, rather than as a trainer. Although capable of mounting armament and bomb racks, the fittings for these were unfortunately not available at the time of acquisition in 1933. The President of Paraguay, Dr. Eusebio Ayala, is known to have flown in the rear cockpit of this aircraft to make several visits to the front.

The aircraft apparently survived the war and continued in service into 1943.

WACO MODEL C

Another of the "mystery" aircraft of the Chaco War, the Paraguayan Fuerzas Aéreas Nacionales acquired at least one Waco cabin biplane in January 1933 from Argentine sources. The aircraft had been bought via a subscription from the employees of the International Products Corporation of Pto. Pinasco in the Chaco.

It seems most likely that the aircraft in question, which received a rather high Paraguayan serial (T-13) was one of the several Model C variants that may have been exported to Argentina. Given the known acquisition date of the aircraft, however, there are only several possible candidates. The sole Paraguayan aircraft is described as having been painted "....overall natural metal or doped aluminum color" with Paraguayan roundels in four wing positions, standard rudder marks and the serial in black characters on the mid-fuselage.

Unfortunately, the Paraguayan Waco was lost in a crash in the Chaco on 3 March 1933 while being flown by Tte. Dentice.

Paraguay apparently acquired another Waco biplane later with a Wright R975E2 engine, but this is believed to have been an ex-Brazilian Model F, and had no Chaco war connections.

CANT 10ter

As cited earlier, when Italian Colonel Colombo was compelled to wind up his successful but short-lived Taxi Aéreo airline in Argentine, he brought the surviving fleet to Paraguay, where he interested the Paraguayan Navy's infant air service in acquiring them.

Besides the Savoia S.59bis "Chaco" covered earlier, Colombo also ordered his crews in Buenos Aires to bring a single Cant 10ter passenger flying boat to Paraguay and this aircraft arrived in either April or July-August 1929, depending on which source is consulted.

Often thought to be a much larger aircraft, this Paraguayan Navy Cant 10ter transport is seen in 1929 at the Sajonia Naval Air Base. A former Argentine registered airliner, the aircraft may have briefly carried the serial H-1 or R-1. An "R" appears in the vertical white portion of the irregularly painted national colors on the tail. (Archivo Cnel. Av. (R.S). A. Pasmor).

Often thought to be a very large flying boat, in fact the Cant 10ter, a 1926 design, had a wing span of only 50 ft 2 in (15.30m) and a length of 36 ft 6 in (11.50m). The single crew member (the pilot) sat in an open cockpit just ahead of and under the radiator of the pusher Lorraine-Dietrich 400hp engine suspended between the wings, and a cabin in the forward hull had seating for up to four passengers with an array of five small windows on either side of the bows.

Only 11 Cant 10ters were built, and the machine that reached Paraguay, c/n 10, had originally carried Italian civil registration I-OLTL, then R-ACVW in Argentina. In Paraguayan Navy service, it had serial H-1.

This unique aircraft had to be withdrawn from service in 1933, right after the arrival of the Macchi 18's, due to a lack of vital spares.

SIKORSKY S-38B

Most sources claim that the three Sikorsky S-38 amphibians acquired by the Bol-Inca Mining Corporation (in other words, Sr. Patino), c/n 114-1 NC303N, c/n 114-10 NC9151 and c/n 114-11 NC9138 had nothing at all to do with the Chaco War.

While this may be true of two of the aircraft, at least one, (probably c/n 114-1) named "Nicolas Suarez," was apparently passed to LAB as early as 24 March 1934 and is known to have carried out at least two flights on behalf of the military into the war zone. Indeed, even German diplomatic records claim that "....as of March 1934, the Government had acquired one Sikorsky S-38".

This aircraft, adorned with high-visibility stripes across the entire width of the upper and lower wings as well as across the breadth of the horizontal tail surfaces, is also known to have had military style rudder stripes on the upper portions of the rudders.

This aircraft survived the war and continued on in LAB commercial service until lost in an accident on 9 October 1941.

Although many historians have stated that the LAB Sikorsky S-38 amphibians acquired starting in early March 1934 were not used during military operations, this is not correct. At least one, named "Nicolas Suarez," was engaged in limited quasi-military support activities, and clearly bears the Bolivian tri-color and high-visibility wing markings in this Cochabamba view. Bolivian Air Force training aircraft are visible in the background. (R.E.G. Davies).

THE BOLIVIAN TRAINERS

CAUDRON C97

Trainers, although acquired in only comparatively small lots by both combatants, played a part in the overall conflict that is seldom appreciated.

It should be noted that, with only limited exceptions, both Bolivia and Paraguay went to war with the pilots who had been trained nationally prior to the commencement of hostilities, and only a relatively small number gained "wings" after full-scale warfare had commenced.

Virtually all of these pilots and crews, however, were trained internally and, to a greater or lesser extent, the quality and durability of the trainers employed dictated the qualifications of the graduates of the respective programs.

It may be generally observed that both nations subscribed to what amounted to the conventional training philosophies that existed in Europe as of about 1928, with a heavy emphasis on the French syllabus. As a result, certain practices crept into the operational experience of both sides that would probably seem foreign to most readers. These included heavy emphasis on rather questionable aerobatics training which, although "pretty" and daring to the observer, had little - if anything - to do with actual air combat. Needless to say, however, these were popular with the cadets of both sides, but took an inevitable toll on both aircraft and trainees. Further, there was very little "crew" training and practically no ordnance delivery or gunnery training, either air-to-ground or air-to-

A stalwart of pre-Chaco War Bolivian flight training, the highly regarded Caudron C 97 was quite a large aircraft for a trainer, and had armament capabilities. This is "Chorolque" at El Alto in 1925. (Ramiro Molina Alanes).

Another view of one of the Bolivian Caudron C 97s, "Oruro I" at El Alto. (Ramiro Molina Alanes).

air. In fact, so far as can be ascertained, no aerial gunnery training of any sort was performed until well after the war was under way by either side, although there was apparently some effort at ground attack training of a rather desultory fashion. Crews were mixed, in the French fashion, so that it was not uncommon to find the pilot of an aircraft an NCO and the observer/gunner an officer or, in other cases, both officers with the gunner out-ranking the pilot. While this system may have worked well for the cosmopolitan French, it resulted in considerable problems for the very rank-conscious military establishments of Paraguay and Bolivia.

Bolivia utilized trainers of six basic types in the formative years leading up to the war and during the war itself. Of these, three were of French manufacture, and one each of English, German and U.S. design. Significantly, only one trainer type was acquired during the formal war period itself, and this only after a rather "hard" sell by the Curtiss-Wright Export Corp.

As may be expected, this mix of equipment led to maintenance and training problems but, it must be stated fairly, although totaling not more than probably 17 aircraft all told (depending on what numbers you prefer), these airplanes, acquired over a 10 year period, served Bolivia out of all proportion to their numbers.

Arguably amongst the most important of these were the Caudron C97s, the first two of an apparent total of five of which were brought to Bolivia by the Swiss Mission consisting of Major Hans Haeberli and Lt. Ernest Meyer in early July 1925, although it has also been reported that at least one other was "in country" as early as August 1924.

Although not the first trainers acquired by Bolivia, these 180hp Hispano-Suiza powered aircraft were the first trainers received which could effectively cope with the rather unfortunately situated Bolivian training field at "El Alto" above La Paz, at an altitude of some 13,500 feet (4,100m).

Major Haeberli reported that the C97s were eminently satisfactory for the conditions found at "El Alto" and that, in fact, the Bolivians themselves referred to the machines as the "tipo Bolivia." During his first six months with these machines, he and his assistant carried out a very respectable total of some 1,500 training flights, and went so far as to conduct a long-range "Good Will" flight to

A close-up of a pre-war scene at El Alto showing a Bolivian pilot about to take his wife for a ride in one of the Caudron C 97s while his mother (and dog) stand by. The man at right is believed to be Major Hans Haeberli, the Swiss officer who single-handedly molded the first generation of Bolivian pilots to fly in the war. Note the unusual left-hinged hatch arrangement of the rear windscreen to permit easier access to the rear cockpit. (Hagedorn Collection).

Buenos Aires (3,000km) with one of them, as well as a dozen other "raids" around the Republic of between 500 to 600kms each.

The C97s, although primarily engaged as trainers, were capable of mounting armament - a feature that endeared them to the Bolivians - and were thus counted as auxiliary combatant types, unlike the other trainers on hand at the time.

By 11 September 1926, one of the first two aircraft had been lost in an accident, but apparently at least three others were acquired, as Bolivian C97s are known to have carried the following names (although, as was the custom of the time, none are known to have had serial numbers): "Illampu," "Chorolque" (later re-named "Cap. Valle"), "Sajama," "Sgto. Flores" (later apparently renamed "Oruro" or "Oruro I") and also possibly "Illimani."

By 12 December 1927, three C97s were still in daily use but only one of the high-time aircraft survived into 1932, but this was used in the war zone (apparently at Villa Montes) to train combat crews into 1934, when it was finally retired due to a complete lack of spares.

Bolivia's C97s were apparently doped aluminum overall and had national roundels in all four wing positions and standard Bolivian rudder stripes. The individual aircraft names were presented in a flowing script on the mid-fuselage atop a cheatline running the length of the mid-fuselage aft from the engine compartment firewall of unknown color while the "Caudron" logo was also carried on the vertical fin on either side in script similar to the individual aircraft names.

CAUDRON C59Et2

At least two reliable diplomatic reports dated between 1 and 15 May 1928 reported that Bolivia had contracted with France for not fewer than 100 Caudron C59Et2 trainers. Like the much earlier Caudron C97s, these were to have been equipped with a 180hp Hispano-Suiza engine (a rather large engine for the C59).

However, apparently this contract was canceled for reasons unknown as no C59s have been positively traced as having arrived in Bolivia. It seems likely that Bolivia had, with the mounting tensions with Paraguay, decided to spend the rather substantial sums involved in this contract on more warlike equipment.

VICKERS TYPE 155 "VENDACE III"

Apparently at about the same time that Bolivia was negotiating for the Caudron C59Et2s noted above (possibly through the agency of the Vickers conglomerate), she was also entertaining the acquisition, as part of the Vickers package of aircraft, of a trio of Vickers Type 155 "Vendace III" trainers.

The "Vendace III" was powered by the 330hp Hispano-Suiza engine, which gave the trainer an outstanding performance at Bolivia's high training field. The Bolivian aircraft were tested on delivery in July 1929 by RAF Flying Officer H. W. R. Banting who apparently was also contracted to perform as an instructor on the machines. The Bolivians were impressed by the aerobatics qualities of the aircraft - even at the high "El Alto" - and mimicked the contortions of Banting at every opportunity. The three aircraft acquired were against Vickers Sales Order No.68-A and cost 9,997.19 Pounds Sterling.

After rendering several hundred hours of training time each, however, the three "Vendace" began to experience engine problems, possibly incident to the rather primitive maintenance care the aircraft were able to receive at the hands of Bolivian ground crews. The manufacturer, perhaps wanting not to point fingers at these rather casual maintenance activities (by British standards of the day) stated that the difficulties were due to the extremely dusty environment at "El Alto." As the engines were the chief problem, the aircraft were grounded individually and collectively at intervals while parts for the Hispano-Suiza engines were awaited from France.

One of the aircraft, No.1 (they were, unusually, given serials "1", "2" and "3" by the Bolivians) suffered the ignominious fate of being rather neatly bisected by the prop of a runaway Vickers "Vespa" at "El Alto" in 1931. At least two unrelated sources suggest that Bolivia may have acquired at least one additional "Vendace", perhaps as a replacement for the wrecked aircraft as, for instance, U.S. Army Air Corps 1LT A. Y. Smith reported seeing four "Vendrome" (sic) trainers at El Alto in January 1932. However, an Order of Battle dated 23 May 1933 showed only two aircraft of this type "in country," although another reliable source claims that a second "Vendace" had been lost to an accident before hostilities commenced.

At any rate, at least one of the "Vendace" trainers served in the war zone, apparently as both a hack and communications aircraft, as far forward as the Saavedra fortress, where it was burned by Bolivian forces in December 1933 prior to their retreat from that area in the face of a determined Paraguayan advance.

No illustrations of any of the "Vendace" trainers have surfaced, and little is known of their coloring or markings.

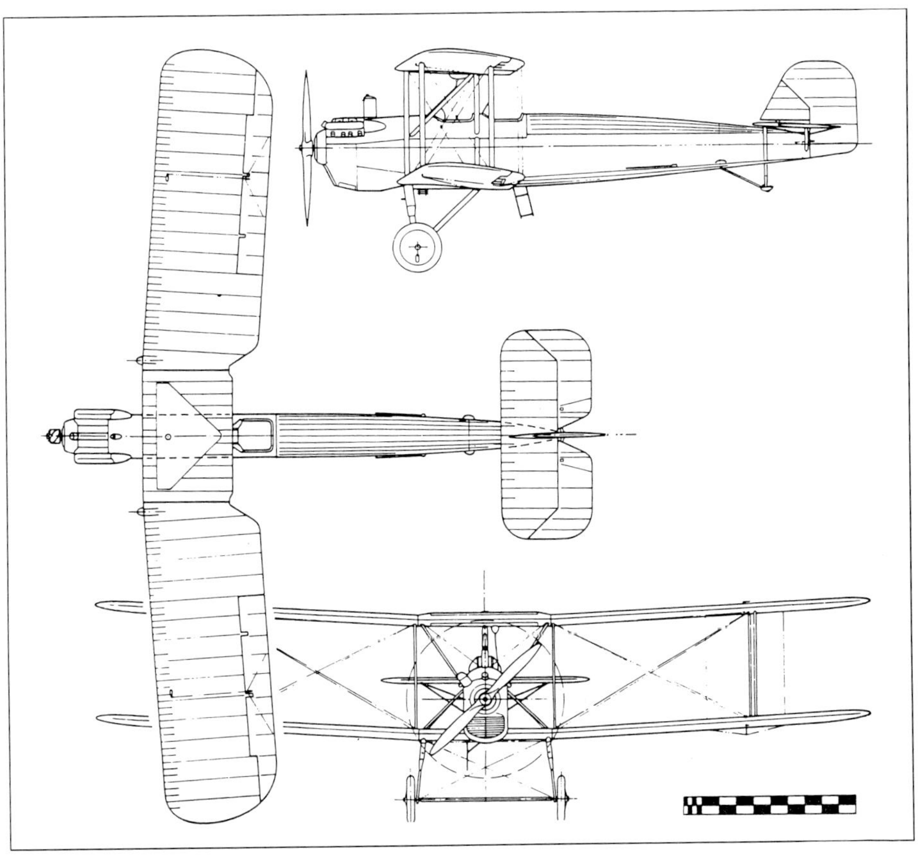

Three-view of another Bolivian stalwart, the Vickers Type 155 "Vendace III" trainer. (Courtesy C. F. Andrews).

MORANE-SAULNIER 35 E.p.-2 AND MORANE-SAULNIER 139

Considerable confusion regarding the variants of the popular Morane-Saulnier parasol trainer series that saw service with both Bolivia and Paraguay has resulted from, basically, mis-identification of the types involved over the years.

In fact, Bolivia acquired four examples of the Morane-Saulnier 35 E.p.-2 (as well as two "Pinguino" taxi trainers), together with the services of the French instructor E. Deckert, via an order placed on behalf of the military by the Ministry of War and Colonization. These aircraft were delivered 22 July 1923 and have also been described in official Bolivian publications as "tipo R.A. Parasols" with 130hp LeRhone or Clerget engines. These were given serials "1" to "4" and rendered good service through at least December 1927, when one was still airworthy.

Other reports that Bolivia received the later but very similar Morane-Saulnier 139 E.p.-2, also with the 130hp Clerget engine, seem completely without foundation, various sources claiming as many as 10 as having been acquired.

The durable M.S.35s featured the typical Morane-Saulnier undercarriage and an adjustable tail skid controlled by the pilot. Not generally known, the aircraft had wing-warp type controls and was marketed as being "inherently stable by design." The aircraft proved rather fragile for "El Alto" conditions, however, but nevertheless provided primary training mounts for most of the pilots who initially went to war against Paraguay and thus have a place in this history.

Its rotary engine blasting a shower of dust at El Alto, it is a wonder that any practical training could be accomplished there on the power available. So highly regarded was the Morane-Saulnier 35 in Bolivia that a postage stamp was minted in its honor! (via Randy Lieberman).

MORANE-SAULNIER
Type 35

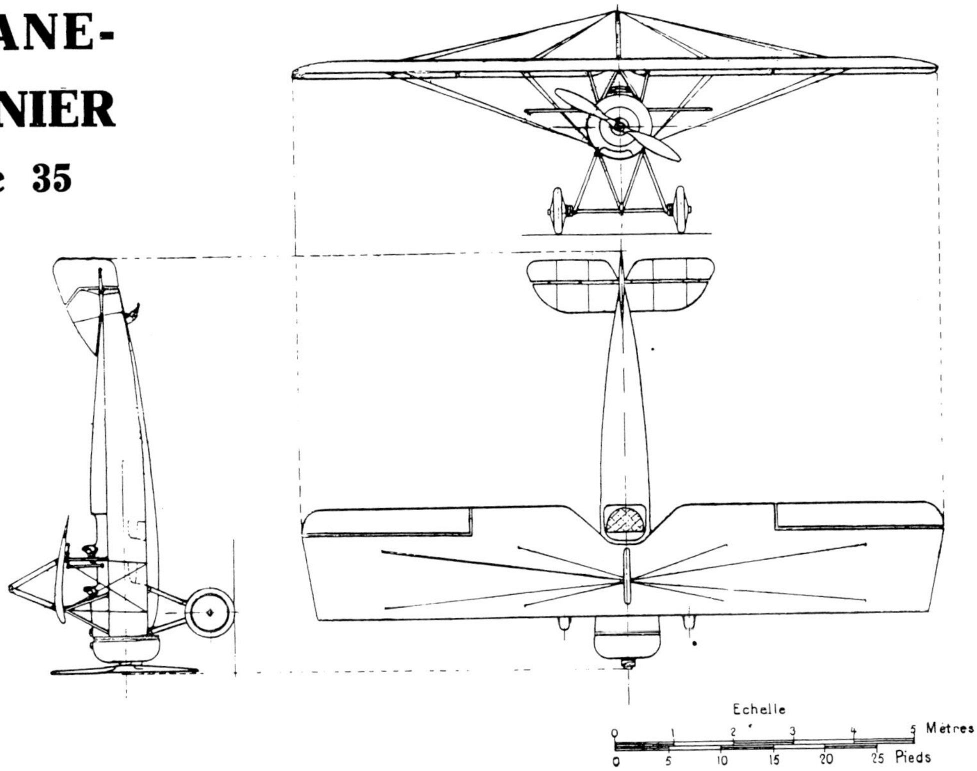

DÉSIGNATION

Morane-Saulnier 35 E.p.-2.
Disposition de la voilure : monoplan.

Moteur "*Gnôme-Rhône*" 80 CV.

DIMENSIONS

Envergure.... $10^{m},565$ Longueur.... $6^{m},767$ Hauteur.... $3^{m},605$

Surfaces

Voilure........................	18^{m^2}
Ailerons........................	$2^{m^2},30$
Gouvernail de direction............	$0^{m^2},45$
Gouvernail de profondeur...........	$0^{m^2},90$

Répartition des poids

		Indices
Poids équipé........................	460^{kg}	0,66
Poids du combustible (essence 70^{l}, huile 17^{l}).	70^{kg}	0,10
Poids utile........................	170^{kg}	0,24
Poids total........................	700^{kg}	1.00

Poids par HP........................	$8^{kg},7$
Poids par mètre carre................	$38^{kg},9$
HP par mètre carré..................	4 HP,4

Particularités et Construction

La voilure est disposée en "parasol".

On peut indifféremment piloter du poste avant ou du poste arrière. Le levier arrière de manœuvre (manche à balai) se démonte facilement. Presque toutes les commandes de vol sont montées sur roulements à billes simples ou à rotules. L'orientation de la béquille est commandée par le pilote en même temps que le gouvernail de direction. Fixé au fuselage par cardan, l'atterrisseur est du type à demi-essieux indépendants.

Destination

Biplace pour école de pilotage (*destiné spécialement au début de l'apprentissage en double commande*).
Peut également être utilisé comme avion biplace de tourisme.

Frequently cited in error, both Bolivia and Paraguay owed much to the pilot tuition provided by the venerable Morane-Saulnier 35 E.p.-2 parasol trainers they acquired. This three-view shows the distinctive planform to excellent advantage ("L'Aeronautique").

CURTISS-WRIGHT CW-16E "TRAINER"

The Curtiss-Wright Export Corporation and its agents in Bolivia had attempted to interest Bolivia in the aircraft they generally referred to as "the Trainer" as early as late September 1933, as part of the total aircraft package with which they virtually re-equipped the Aviación Militar.

Both Curtiss-Wright and the Bolivians realized that there was a rapidly developing need for capable training aircraft by that time, as only single examples of the previously acquired trainers remained airworthy at any one time by late 1933, and newly commissioned cadets were receiving, at best, a rather hesitant and patch-work tuition.

In fact, the Bolivians were actively considering contracting out the pilot training responsibilities to German cadre (due to the influence still very pervasive of General Kundt). However, this came to naught.

"Cliff" Travis, seizing upon this decision, once again opened the question of CW-16Es for Bolivia in late September 1933, and in fact queried the home office on behalf of the Bolivians, asking whether or not it would be possible to "mount a larger engine on the aircraft for use at La Paz." He voiced his concerns that "the standard 165hp job could probably not get off the ground up here with both a student and instructor on board" and instead suggested that the factory investigate whether it would be possible to mount the larger "suped-up seven cylinder engine" on the airframe, although it is not clear what engine he had in mind (the 220hp Wright engine seems likely).

By that time, of course, virtually all flying training had been transferred from La Paz to Villa Montes itself, so the abilities of the standard 165hp R-540E (Wright J-6-9 engine) were not a problem. With this in mind, Travis finally received Bolivian orders for three

Very rarely illustrated, this is one of the trio of Curtiss-Wright CW-16E trainers acquired by Bolivia in late January 1934 to replace its rather exhausted pre-war trainers. They were actually deployed and training was conducted at Villa Montes near the front. (Ramiro Molina Alanes).

CW-16Es with the R-540E engine at a bargain $6,000.00 each c.i.f. Arica. These were shipped from the St. Louis factory by mid-December 1933, arriving finally at La Paz on 26 January 1934, taking "Cliff" Travis somewhat by surprise, as he had not expected them quite so quickly. They were quickly trans-shipped to Villa Montes, by which time he reported that "they are anxious to get them [there] quickly." He reported that he thought they might be able to sell two more CW-16Es to Bolivia on the basis of the evaluation and approval they received at Villa Montes but this did not happen. Travis stated bluntly that he was "pushing hard on this because this would make all the new pilots think `Curtiss-Wright', which is bound to have influence on future purchases."

One of the CW-16Es (No.3) suffered an accident at Villa Montes at the hands of a student on 19 May 1934 but this was apparently repairable, even though the report at the time stated that the aircraft "was a write-off."

The other two aircraft continued on in intensive training activities at Villa Montes. However, by Christmas 1934, both were grounded due to damage to their props, and were apparently not returned to service, except one briefly, before the Armistice.

The surviving pair of CW-16Es remained in FAB service as late as 4 September 1944, by which time they were assigned to the 2° Escuadron de Entrenamiento at comparatively low-altitude Cochabamba, but both were listed as "unserviceable due to parts." Both were amongst 17 former FAB training aircraft turned over to Government sponsored civil flying schools sometime between September 1944 and September 1945. As mentioned earlier in the case of the surviving Osprey's and Hawk's, however, the Bolivian Government formally requested allocation of unused Lend-Lease funds on 19 March 1945 to effect repairs on three "Curtiss Trainers" and there is little doubt that these were the CW-16Es.

No details on the coloring or markings of Bolivia's trio of CW-16Es are known.

JUNKERS A.50 "JUNIOR"

Paraguay was not alone in employing unusual aircraft. Bolivia's hard-pressed training establishment enlisted three LAB Junkers A.50 "Junior" trainers around May 1933 one of them, "Piray" seen here undergoing overhaul with Junkers W34 s/n 101 in the background ("Condor").

Mounting 80hp Armstrong-Siddeley "Genet" engines, LAB had acquired three of these sturdy ab initio trainers for use at its Cochabamba base some time in 1930.

As the conflict with Paraguay expanded and Bolivia's trainer establishment dwindled, the three A.50s were finally incorporated into the Aviación Militar entirely in June 1933, being taken to the Muñoz air base in the Chaco where rather desultory primary training was undertaken on them.

The aircraft continued bearing their LAB names: "Warnes," "Acre" and "Piray". Of these, "Acre" was lost in a training accident some time in 1933 while "Piray" was eventually transferred to Roboré on the railroad between Puerto Suarez and Santa Cruz while "Warnes" was retained at the advanced Puerto Suarez base itself until the end of the war. The surviving pair were returned to LAB after the war.

THE PARAGUAYAN TRAINERS

FLEET MODEL 2

The most important dedicated training aircraft used by Paraguay during the Chaco War were, unquestionably, five Fleet biplanes (sometimes given as Consolidated-Fleet) purchased in Argentina in 1931.

These aircraft, however, are amongst the several other puzzling aircraft of this little known conflict. While the five aircraft purchased were, almost unquestionably, Fleet Model 2s, known illustrations of these airplanes suggests that they have some characteristics of Fleet Model 3s and Model 10s.

Of the five aircraft acquired, four had the 125hp Kinner B-5 engine while the other had a larger 160hp engine. These received Paraguayan serial codes E.7, E.9, E.11, E.13 and E.15 and, of these, E.13 crashed in 1935 and was written-off (probably at the hands of Tte. Antonio Esteche). The others seem to have survived into January 1949, by which time at least three had passed to LATN (one of which was the 160hp powered variant). One survives to this day and has been restored, although unfortunately in non-authentic markings, the sole surviving veteran of the Chaco War.

The sole surviving aircraft of the Chaco War, this recently restored Fleet Model 2 is unfortunately not painted in genuine wartime colors. It was photographed at Ñu-Guazu Air Force Base in December 1991 (Antonio Sapienza).

Another trainer, this Paraguayan Fleet Model 2 survived the war to enter service with the Army sponsored airline, L.A.T.N. The source of these aircraft is unknown, but was probably either Argentina or Brazil (Archivo Cnel. Av. (R.S). A. Pasmor).

This otherwise innocuous photo of a young Paraguayan pilot in front of a Fleet Model 2 is remarkable in that it was taken in the war zone (Isla Poí) in 1933 and appears to have Paraguayan national colors painted in large bands like invasion stripes under the inboard wings! (Archivo Cnel. Av. (R.S). A. Pasmor).

Left: Paraguayan Fleet Model 2 s/n E.13, obviously bearing camouflage, at a most unlikely location - the forward Isla Poí aerodrome in 1933. Tte. Agustin Pasmor, chief of maintenance, gives scale to the small aircraft. (Museo de Historia Militar via A. L. Sapienza).

Like Bolivia, Paraguay also utilized the Morane-Saulnier 35 E.p.-2, and a very handsomely turned out example is shown here (coded E.4 in small characters) in full markings at Ñu-Guazu in 1932. (Archivo Cnel. Av. (S.R). A. Pasmor).

Another of the Paraguayan Morane-Saulnier 139E.p.2 trainers that served so well and faithfully, this one is coded E.5, and is departing on a training flight at Ñu-Guazu in 1932. (Museo de Historia Militar via A. L. Sapienza).

This lineup of Paraguayan pilots, with their French advisor, pose before the Morane-Saulnier 35 E.p.-2 on which they all took tuition. Many went on to glory in the Chaco. Incredible as it may seem, the M.S.35 employed wing-warping control! (courtesy Georg von Roitberg).

A close-up view of Paraguayan Morane-Saulnier 35E.p.-2 s/n E.5 at Campo Grande in 1928. Aboard are 1° Tte. Leandro Aponte and, in the rear, 2° Tte. Emilio Rocholl, the first Paraguayan pilot lost in action against Bolivia. (Collection of A. L. Sapienza).

CANT 26

In its search for virtually any useful airworthy aircraft, Paraguay found (via Sgto. Nicolas Bó) one rare Cant 26 biplane available for sale at a reasonable price in Argentina.

A de Havilland "Moth" look-alike, the Cant 26 had an 85hp ADC "Cirrus II" engine, and had carried Argentine civil registration R-183. In Paraguayan service, it gained the nominal "transport" serial T-6 and had entered service some time in 1932.

Unfortunately, the aircraft was lost in an accident on 5 May 1933 killing its crew as well, Cpt. José D. Jara and Tte. Niemman.

The parade of unique types goes on. Paraguay bought one Cantieri Cant 26 in Argentina (the former R-183) and coded it in the transport category as T-6. A de Havilland D.H.60 look-alike, this aircraft has never before been illustrated. It was lost on 5 May 1933. (Hagedorn Collection).

SOCIETA ANONIMA MECCANICA LOMBARDA S.A.M.L. A.3 (S-1)

Left-overs from the May 1922 revolution, at least two S.A.M.L. A.3s (also sometimes given as S-1s) remained on hand for service with Paraguay by 1927.

These, eventually coded M.E.T.1 and M.E.T.2, rendered only incidental training support to the more efficacious Morane-Saulnier, Hanriot and, eventually, Fleet trainers, being employed for the most part as "roulers," although one apparently had enough airworthiness left in it to crash in 1928.

The other survived through the war years, and appears to have been jointly used by the Army and Navy to train pilots, being passed eventually to the Aero Club del Paraguay in 1937. However, it would appear that this aircraft that survived may in fact have been a composite of the second of the original pair mated with an engineless S.A.M.L. purchased from the redoubtable Sgto. Nicolas Bó in 1929. As far as can be determined, all these aircraft had the 130hp Colombo water-cooled engine.

5

Things That Didn't Happen

As in any conflict in which the combatants had little or no indigenous manufacturing capability, and in which the rush to war may be said to have overtaken adequate preparation, the Chaco War and, in particular, the aviation aspects of that war, have given rise to a number of myths, half-truths and rumors amongst historians of the subject since the earliest days after the armistice.

Some of these "non-events" can almost certainly be attributed to incomplete information, as some of the arms bazaar activities engaged in by both sides have remained, perforce, secret even to this day, and may never be known with clarity.

However, the authors of this work feel compelled to include at this juncture at least a summary of some of these otherwise unaccountable aircraft.

We have elected the convention of recording these, in the following pages, in alphabetical order based on the manufacturer usually cited, and have included such information as has come to the surface in each instance.

One of the truly little-known aircraft purchases contemplated by Bolivia as of October 1929 was for five Andiz Monoplanes. Designed and marketed by Andiz, Incorporated under the leader-

Although it is not at all clear what they would have done with them, Bolivia made overtures to acquire as many as five of the odd Andiz Model 2-60 "Monoplane" trainer. The deal apparently fell through when the manufacturer went belly-up in the early days of the depression (Andiz).

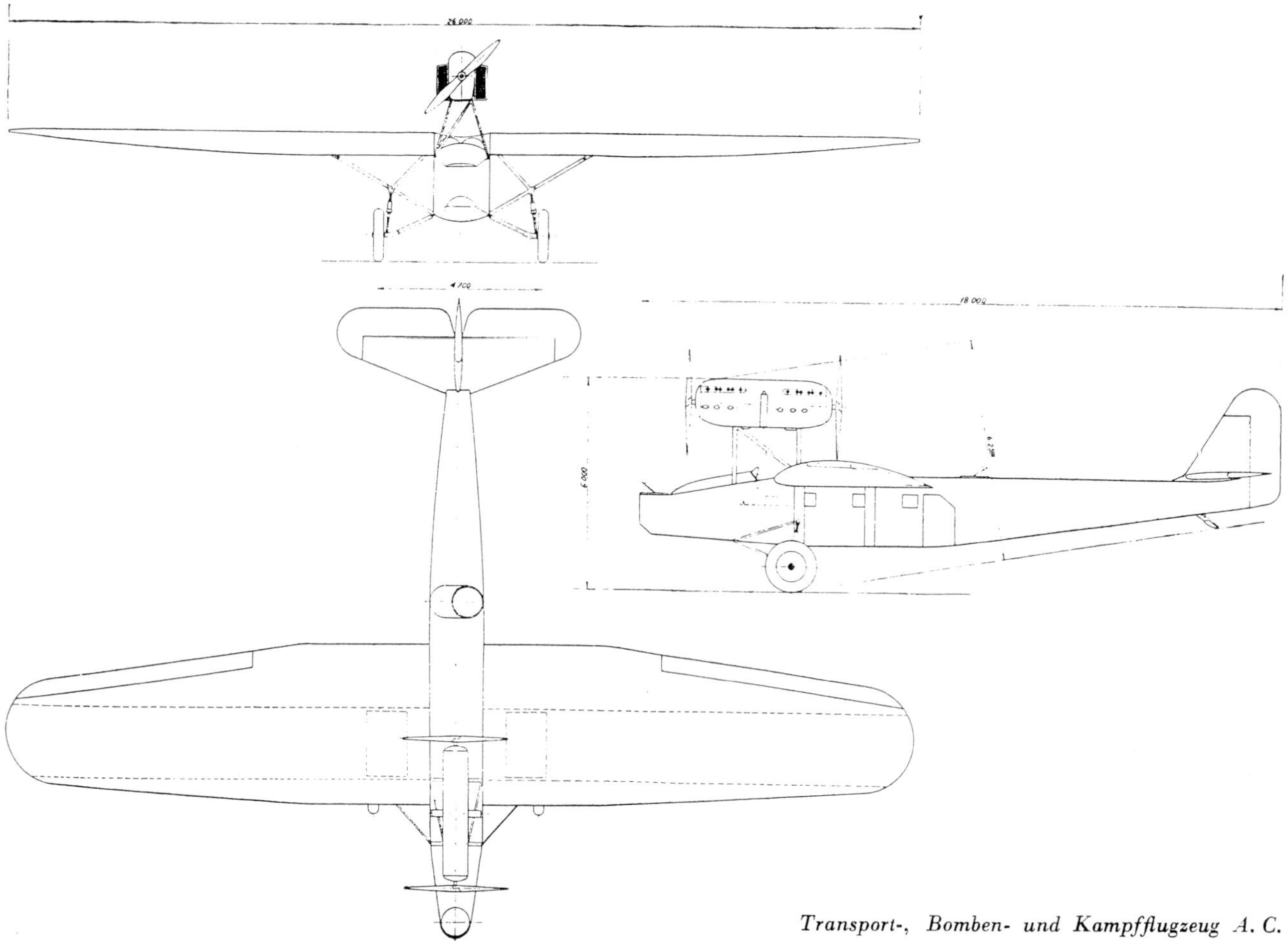

Things that might have been.....included this three-view of the huge Comte A.C.3 bomber transport, at least three of which were ordered by Bolivia ("Schweizer Aero-Revue").

ship of W. H. H. Anderson, an MIT graduate, this rather frail looking parasol monoplane was to have mounted either a 60 or 80hp "New Series" Anzani engine (although designed to mount any engine up to 130hp), and the first Andiz gained a U.S. experimental license, X-4244. Apparently this order was canceled when the Andiz firm went belly-up during the Depression, although even had these aircraft been delivered, it is hard to imagine how Bolivia would have employed them at "El Alto."

Several sources cite Bolivia as having operated a mix of Monosoupape, LeRhone, "Lynx" and "Viper" powered variants of the Avro 504 biplane trainer circa 1926, a total of not less than 11 being quoted in two otherwise reputable sources. One well known historian claims that at least a few of these were ex-Peruvian machines. However, official Bolivian histories of aviation in that country make no mention whatsoever of this basic type nor have any intelligence reports surfaced that support the claims.

Paraguay evinced interest at one point in acquiring a number of ex-Argentine military Bristol F.2B "Fighters" some numbers of which had been constructed in Argentina by the Fabrica Militar de Aviones. Argentina was still operating at least 10 of these as late as July 1931. The deal was frustrated for unknown reasons, however.

One Bolivian order that very nearly came to fruition, however, could have had grave consequences for the populace of Asunción. These were the strange looking trio of Comte A.C.3 twin-tandem-engined heavy bomber/transports ordered from that Swiss manufacturer (through one Ph. Vacano, at the instigation of Major Hans Haeberli) around May 1930. Bearing c/n's 27, 28 and 29, apparently only one of these was actually completed, although it was painted in Bolivian markings while being readied for delivery. However, a change in government in Bolivia resulted in the contested cancellation of this 110,500 Boliviano order. To have mounted two 600hp Hispano-Suiza engines and with nose and dorsal gun positions (and provision for a ventral gun as well) the exact bomb load of which this aircraft was capable is not clear, nor is its maximum range. However, it seems clear what Bolivia had in mind for the A.C.3s.

As mentioned earlier in this account, Curtiss-Wright Export Corporation representatives had made at least an attempt to sell as many as nine Curtiss-Wright A14R "Ospreys" to Paraguay, at about the same time as it was marketing the similar C14R light combat aircraft to Bolivia. This arrangement appears to have been frustrated by Paraguay's inability to meet Curtiss-Wright's terms at the time, however. At about the same time (circa 15 February 1933), a Curtiss-Wright representative named James H. ("Jim") Spencer was in Santiago attempting to interest the Paraguayan Government in solitary examples of the Curtiss D-12 (in-line) powered Curtiss "Falcon" and the Curtiss "Hawk". Paraguay was apparently particularly interested in the "Hawk," for which Spencer was negotiating with the Paraguayan Minister in Santiago "and also through an official of the same Government in Buenos Aires," who made a counter-offer for the aircraft. Stipulating that any deal must include "full armament and bomb racks" the Paraguayan offer was "so ridiculously low" against the asking price of $20,000 (armed) and $15,000 (unarmed) that even the Curtiss-Wright entrepreneurs turned the offer down. The "Falcon," often referred to in Curtiss-Export correspondence as "The Wooten "Falcon," (named for a pilot who demonstrated that solitary aircraft extensively) and the "Hawk" were both, essentially, left-overs from the disastrous Curtiss attempt to establish a "factory" at Los Cerrillos near Santiago, Chile. Eventually, however, these two aircraft, after exceedingly complex business and diplomatic contortions, ended up being sold to the equally aircraft-hungry Government of Peru. It will be recalled that Curtiss-Wright had also attempted to interest Bolivia in these same two aircraft and, while Bolivia came closer to the contract price than Paraguayan offers, they insisted on terms only if Chile would concede to allow the two aircraft to be flown out of Chile to Bolivia, obviously a political gambit to set a precedent in favor of Chile permitting arms transfers to Bolivia.

Certainly the most provocative still-born aircraft acquisition deal of the entire war, however, was a Bolivian follow-up to the aborted Comte A.C.3 heavy, long-range bomber order. This, of course, was the attempt to acquire four Curtiss-Wright "Condor" bomber/transports, with the avowed intention of using them in bombing attempts on Asunción itself.

The very first known Bolivian interest in acquiring heavy, long-range bombers after hostilities commenced came around mid-October 1933, when the Curtiss-Wright representative in La Paz, after a very late dinner with several high officials of the Government, stayed up late to draft a letter to the home office in which he stated that "....the Government wishes to acquire 10 large bombing planes, and it is a choice between Junkers and ourselves." The Government's chief supporter of Curtiss-Wright, Lopez, armed with data supplied him by Curtiss-Wright, managed to sway other key officials towards the "Condor" choice and, by 8 November 1933,the terms of payment for a "prospective" order were being negotiated. The total number of aircraft involved was still 10, but of these, "at least seven would be from Curtiss-Wright, provided we can arrive at a mutually agreeable deal. They [the Bolivians] will spend from $800,000 to $1,000,000 with us if we can get together. Their position on this is: pay 25 or 30% cash, with the contract, and the balance in equal monthly installments over a period of 18 months - the entire contract to be guaranteed by the Banco Central." The Banco Central, while not a Government bank per se, held all of the Governments Gold reserves, as well as "Miners Drafts" (which were really monthly taxes from the all important tin mines) totaling some

At least one of the Comte A.C.3s was actually finished and painted in Bolivian markings. (Hagedorn Collection).

CURTISS - WRIGHT CONDOR BOMBER BT-32

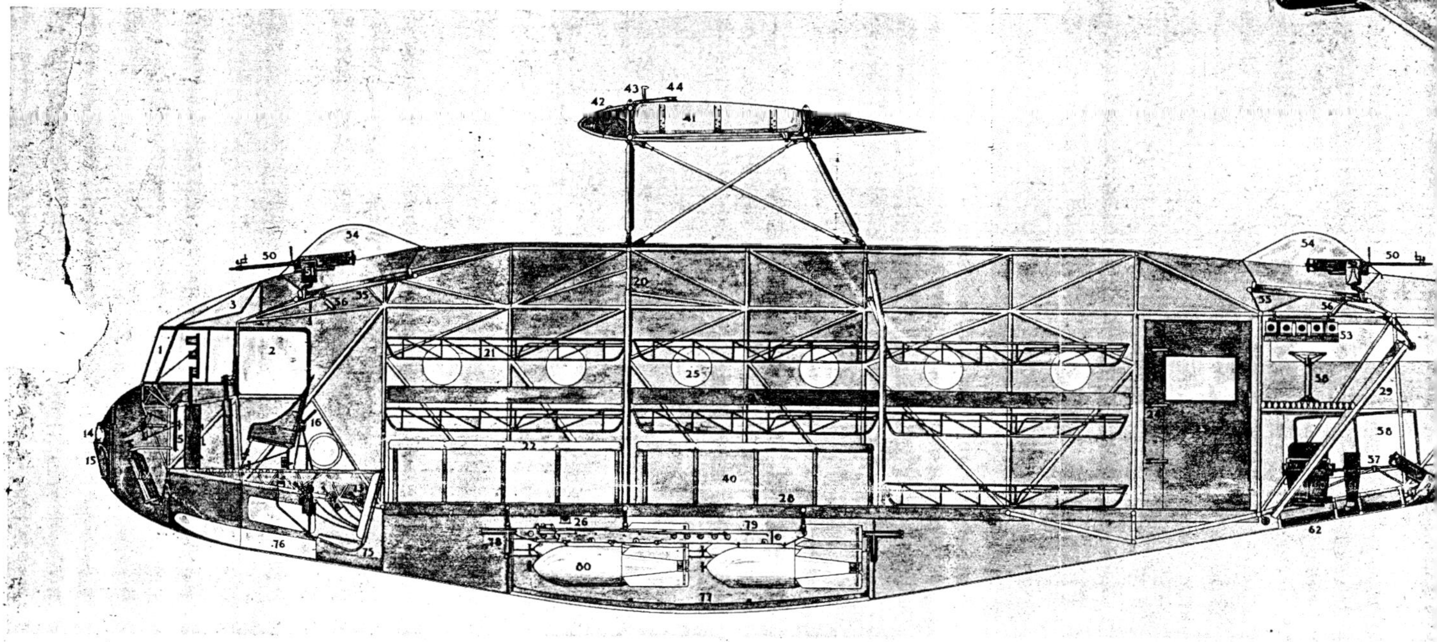

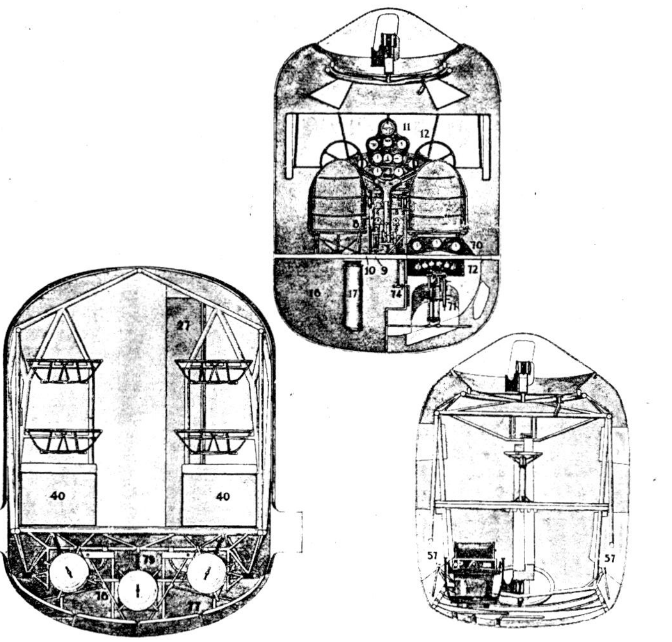

● PILOT'S COMPARTMENT

1. Windshield
2. Sliding Window
3. Amber Glass Panel
4. Dual Flight Controls
5. Adjustable Rudder Pedals
6. Adjustable Pilot's Seat
7. Engine Control Unit
8. Stabilizer Control
9. Rudder Tab Control
10. Elevator Tab Control
11. Flight Instrument Panel
12. Engine Instrument Panel
13. Fire Extinguisher
14. Red Fog Light
15. Landing Lights
16. Landing Gear Hand Retracting Crank
17. Flares
18. Baggage Compartment

● FUSELAGE

20. Welded Steel Tube Structure
21. Ambulance Litters
22. Seat Cushions
23. Cabin Door
24. Emergency Door Release Handles
25. Windows
26. Fuselage Jack Pads
27. Lavatory Door
28. Cabin Floor
29. Control Cable Guard

● TAIL

30. Welded Steel Tube Structure
31. Duralumin Ribs
32. Duralumin Rudder Tab
33. Navigation Light
34. Tail Wheel
35. Long Stroke Oleo
36. Tail Wheel Fairing
37. Stabilizer Adjusting Screw
38. Stabilizer Adjusting Frame

● FUEL SYSTEM

40. Fuselage Gasoline Tanks 4 @ 112½ U. S. Gallons
41. Wing Gasoline Tanks 4 @ 75 U. S. Gallons
42. Gas Tank Filler Caps and Manifold
43. Fuel Tank Vent
44. Electric Fuel Gauge—Tank Unit

● MACHINE GUN INSTALLATION

50. Machine Gun—Colt Browning
51. Ammunion Box Holder
52. Link Collector Bag
53. Ammunition Storage Rack
54. Gunner's Turret
55. Fixed Ring
56. Hand Release Lever
57. Auxiliary Gun Mounting Socket
58. Gunner's Seat
59. Sliding Window

● CAMERA

60. Fairchild
61. Vertical
62. Sliding D

● BOMBING

70. Bombard
71. Estoppey
72. Bomb Re
73. Bomb Ar
74. Bomb Do
75. Bombard
76. Glass Wi
77. Bomb Do
78. Bomb Do Mecha
79. Bomb Sh
80. Bomb—

● COMBINA

I 6 30
II 3 60
III 2 110
IV 20 12
V 6 30 Bo
VI 3 60 Bo
VII 2 110 Bo

Bolivia had, by 1935, pinned many of its hopes for a successful outcome to the war on delivery of up to four Curtiss-Wright BT/AT-32 "Condor" bomber/transports. This extremely rare cutaway shows how the aircraft would have been configured (Curtiss).

$200,000 per month. These payments, in essence, were the means for financing the very lucrative "Condor" purchase.

The Curtiss-Wright representative quoted the Bolivians an "air-delivery" price of $70,000 each for the "Condors" - which included an `extra commission' to smooth the transaction. The Curtiss rep on this deal, Owen Shannon, went on to report that "I was also asked to secure about eight good U.S. Army bombing pilots who might be available and out of jobs and looking for a little excitement and money, to come down with the ships and go to work in the Chaco for the Government. This was a confidential request, so do not mention it even to the Consul General."

By 15 December 1933, the new Bolivian Minister of Defense, Benavidez, and the new Minister of War, Jose Antonio Quiroga, had drawn up a list of new aircraft acquisitions to support the war effort in the Chaco. Amongst these Owen Shannon reported that "we have managed to place five "Condors," but there is talk by the Estado Mayor of increasing this quantity to nine units," although this fell somewhat short of Shannon's 6 November 1933 pronouncement that "my one great job [here] is to sell from three to 10 "Condors." As the League of Nations Commission was in La Paz at the time, the powers-that-be were hedging their bets, however, as no one knew for certain if the war would last much longer or not. However, the general feeling was that the war would indeed drag on and the sitting government were making financial plans accordingly. The plan had also shifted subtly to indicate that Patino Mines "or even Simon I. Patino himself" would guarantee the credit for the sale of the five "Condors" for Bolivia. Curtiss-Wright Export, however, rather bluntly stated privately amongst their own hierarchy that "naturally, our aim is to give them as little credit as possible and with the safest guarantees." However, they also observed that "there is absolutely no doubt that things are moving in the right direction for us and, if they do buy anything, our "Condors" will be bought."

C. W. Webster, President of Curtiss-Wright Export, was heavily involved in this high-stakes gamble. In early 1934, the Government demanded lower prices for the "Condors," by which time the ability of Curtiss-Wright to deliver against the proposed delivery dates had been lost, since the aircraft on the production line intended for Bolivia had passed to another customer with ready cash, Colombia. Originally offering delivery in 75 days from date of contract, this was later modified to 45 days. Then the Government insisted that the aircraft be "delivery inspected" by, of all agencies, the Bureau Veritas. Curtiss-Wright was apparently completely ignorant of this organization, but, after inquiries, agreed to this condition. However, after learning that it was in actuality a French organization which (in the words of Webster) "is not so good," he went on to state that "Bureau Veritas inspection of the "Condors" before they leave for La Paz is the most ridiculous thing I have ever heard of. It is merely useless expense and a nuisance. They probably never saw a "Condor" before and what good can be gained by having their man in St. Louis? Haven't we delivered enough airplanes to convince the Government that we are more to be depended upon than some outside party? Does the Government believe for one minute that we would start a bunch of American pilots, all friends of ours, on a 6,000 mile flight unless the ships were O.K.? We certainly do not intend to permit them to come into our factory and instruct us in what goes into the manufacture of a "Condor!""

Webster also volunteered, while in La Paz, to "offer my own services gratis and, as a friendly gesture, to handle the delivery of the "Condors." I also agreed at the request of the Government to secure some pilots, who would remain in Bolivia and handle them in service. I told the Government that it would cost from $150 to $200 per week for good pilots. I can now secure them as I cabled you for $600 for one Chief Pilot, $500 for three first pilots and $400 for Co-pilots. The Government now decides that these men have to receive 30% of their salaries in National currency. All the `good' pilots want American dollars and will not accept any part of their salaries in National currency." Curtiss-Wright had apparently learned some hard lessons as a result of similar involvements in Colombia and China.

Webster was by this point obviously getting exasperated over the entire "Condor" order, especially in view of the legal risks he was taking, personally, to achieve the sale. He went on regarding the costs for the aircraft delivery: "While in La Paz, I told the Government that it would cost approximately $8,000 per aircraft for insurance and flight expenses. It will not be less than that. I also asked the Government to transfer $32,000 to New York to apply against those costs. I do not care if they send me the money or send it to the Bolivian Consul. In fact, I do not care if they send it at all, but if they wish to get these "Condors" to La Paz, they should send it to someone in New York, and quickly."

In fact, Webster applied for U.S. Civil Registrations for the, by now, four aircraft deal, in his own name. These aircraft were as follows:

Model	Constructors Number	U.S. Registration
BT32	59	NC11729 (later NR11729)
AT32C	60	NR11730
AT32C	61	NR11731
AT32C	62	NR11732

Bolivia did, in fact, place the long-awaited order before the adoption by the U.S. Congress of the Joint Resolution of 28 May 1934, making illegal the sale of arms and munitions of war to the governments of nations at war in the Chaco. Needless to say, this placed Curtiss-Wright Export and Webster in a most unenviable position, as by then a contract had been signed and monies exchanged between the parties.

The four "Condors," although bearing slightly differing Model designations, were indeed configured as bombers. The racks on the wings alone could bear bombs weighing between 30 to 120 pounds each, while the main internal bomb bays could accommodate 300 to 1,100 pound weapons. They also had provisions, in rather odd manually controlled turrets and hatches, for up to five .30 caliber machine guns. As heavy bombers, they were guaranteed to cruise at 265km/hr at 4,000m to a range of 725km. As armed reconnaissance aircraft, with a minimal bomb load, they were capable of not less than 2,200km. This latter capability was apparently what interested the Bolivians the most.

With the passage of the Joint Resolution, however, Curtiss-Wright was handed yet another obstacle to the delivery of the four aircraft. In an intrigue nearly 12 months long that is still not known to its full extent, the four aircraft were stripped of their bomb racks, turrets and other military equipment and, bearing the U.S. civil registrations noted above, were nominally "sold" to a "new" airline called "Tampa-New Orleans-Tampico Airlines, Inc." which immediately applied to fly all four, in formation, to (of all places) Arica, Chile, on the understanding that they were to be engaged there in "a commercial enterprise." Apparently with the collusion of key U.S. officials, this most unlikely flight did, in

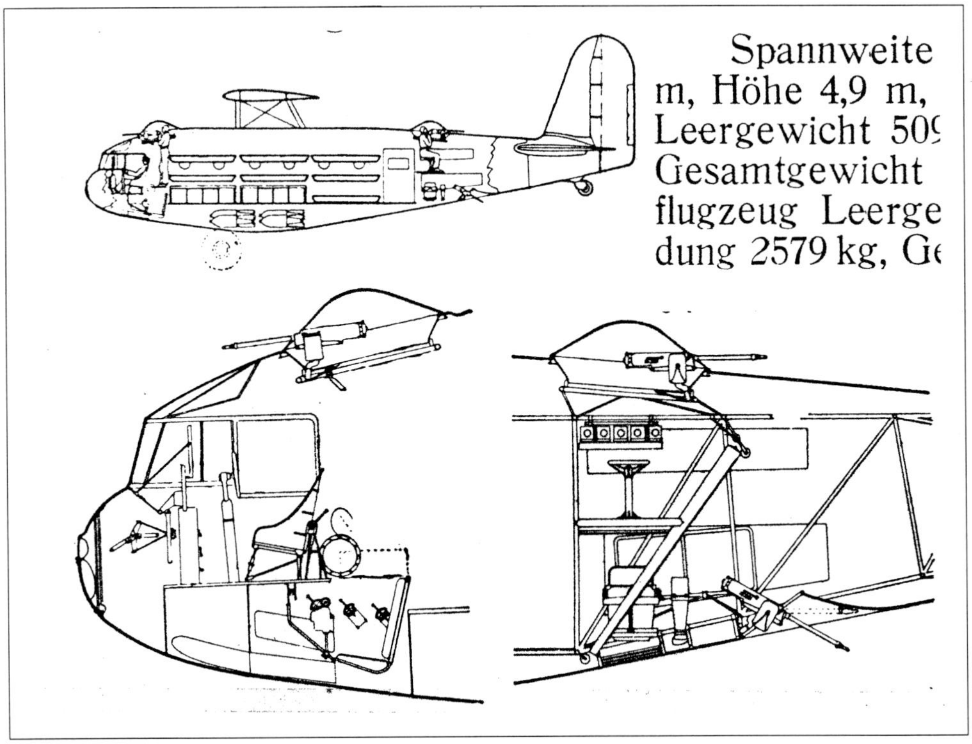

Part of the undoing of the Bolivian Curtiss-Wright "Condor" purchase was the intended gun turrets and guns for the aircraft, which have seldom been illustrated. The turrets were all hand-operated and mounted .30 caliber weapons. (Hagedorn Collection).

fact take off, and the aircraft were flown from Robertson, Missouri (where they were built) to Patterson, Louisiana and from there on to Brownsville, Texas and via intermediate stops to Lima, Peru. In the meantime, the press had become aware of this boondoggle and blew the whistle. The U. S. Department of Commerce requested the Peruvian Government to seize the aircraft on the grounds that they had "violated the course outlined in the flight permission" it had issued and, upon arrival in Lima, the DoC canceled the aircraft registrations.

By this time, April 1935, the Chaco War was very nearly over, and Bolivia decided not to make an issue of the aircraft and, essentially, took the loss. The U.S. Attorney General, meanwhile, sent a Special Assistant to Lima to question the U.S. crews of these aircraft and, following their candid testimony, brought indictments against Curtiss-Wright Export Corp., Curtiss-Wright Aeroplane and Motor Co., the Barr Shipping Co., John S. Allard (by then President of C-W Export), Clarence W. Webster (by then C-W Sales Agent for Latin America) and Robert R. Barr - not for the attempted "illegal" export of what were, essentially "commercial transports" to Bolivia, but for the smuggling of the 15 machine guns and turrets to Valparaiso, Chile, intended to reconvert these aircraft to bombers! Also indicted were Cliff K. Travis and Heustis I. Wells, the latter the "Commanding Officer" of the four "Condors" flown to Lima. A similar indictment was lodged against the same defendants for attempting to sell a Curtiss "Cyclone Falcon" and two Curtiss "Hawk IIs" to Bolivia (other reports claim the shipment consisted of three "Falcons" and two "Hawks"), these aircraft being the examples cited earlier in this account which seem to have simply disappeared.

Ironically, all four of the "Condors" eventually gained Peruvian civil registrations (OA-FFA-272, OA-FFB-273, OA-FFC-274 and OA-FFD-275) and were flown by Condor Peruana de Aviación S.A. commencing around February 1937 as freighters. At least one of these eventually ended up, as a transport, with the Peruvian Air Force.

A myth surrounding an aircraft alleged to have been acquired by Paraguay for service during the war involved the purchase, from Argentine civil sources, of a single de Havilland D.H.80 "Puss Moth". As a matter of fact, there was an element of truth to this purchase: a D.H.80 was bought, but delivery was barred when the Bolivian ambassador in Buenos Aires made the Argentine Government aware of this (and other purchases) and the Argentine Government promptly impounded the aircraft. As there were only four known D.H.80s in Argentine at this point, this aircraft was probably either R-193, R-194, R-200 or R-209.

Hard-pressed for quality combatant types circa 1929, Paraguay made at least two known attempts to acquire quantities of the Argentine license-built Dewoitine D.21 C-1 monoplane fighter. Built in Cordoba by the Fabrica Militar de Aviones, the 1929 attempt revolved around between three to five of these state-of-the-art fighters. A second attempt, in 1934, involved as many as 15 of the aircraft, but again Paraguay was frustrated, apparently due to the unwillingness of Argentine authorities (although sympathetic to Paraguay) to fly in the face of League of Nations sanctions on arms supply to the belligerents.

In a little-known transaction, in which money had actually changed hands, Paraguay purchased five Fokker C Vd reconnaissance-bombers in Holland and, according to Paraguayan historian Alfredo Seiferheld in his Economia y Petroleo durante la Guerra del Chaco these aircraft were actually being loaded aboard a ship in Holland when the deal was "discovered" by Bolivian agents. Since there was an immediate complaint by the Bolivian Government, Fokker suspended the operation and the aircraft seemingly disappeared into the void.

Another player aspiring to secure aircraft orders from Bolivia during the war was none other than the U.S.-based Fokker Aircraft Corporation of America and its subsidiary Atlantic Aircraft Corporation of Teterboro Airport, New Jersey. As early as September 1933, sniffing the possibility of lucrative export orders to the belligerents, Fokker is known to have offered a 10% sales "commission" to key Bolivian officials on sales of aircraft and up to 20% on spares, if business could be thrown their way. The types of aircraft involved or suggested have not been discovered, unfortunately, but may have included a variant of the Fokker YO-27/B-8 monoplane twin engined bomber supplied in small numbers (six each) to the U.S. Army Air Corps in the very early 1930s. Fokker is also known, like United Aircraft Corp., to have offered the sale of Curtiss aircraft with "Cyclone F-3 engines, six pursuit and six observation planes, at cut-throat prices." The reader, if confused by all of this, should bear in mind that Curtiss itself would sell to any broker (including its own "sister" organization, Curtiss-Wright Export Corporation) provided the price was right. The airframe manufacturer wasn't terribly concerned with who the ultimate customer might be, although this would seem to be rather counter-productive. Indeed, officers of Curtiss-Wright Export, upon learning of Tony Fokker's ploy, commented that "he had attempted to do this same stunt with us in Argentina, but we managed to beat him to it."

Unrelated directly to Tony Fokker himself, however, was another aircraft acquisition that concerned Curtiss-Wright Export greatly. It seems that as of 18 July 1933, the Bolivian Consul in the U.S., Sr. Decker, had learned that three former American Airways Fokker F-XA tri-motor transports (and a large number of spare engines) were being offered for sale rather inexpensively (the lot for $30,000) at Chicago. In fact, the Bolivian Government went so far as to instruct Sr. Decker to send someone to Chicago to inspect these airplanes and to see if they could be converted to bombers. Apparently C. W. Webster, then in New York, managed to convince Sr. Decker that "this would be a bad deal, as the ships had been built in 1929 and had had a hell of a lot of service." Webster pointed out to Decker that, although the purchase price of $30,000 seemed like a good deal, that this would be "only the beginning of expense, as the aircraft would have to be completely overhauled, crated and

Here are the four Bolivian "Condors" as they appeared shortly after arrival at Lima, Peru, where they were interned. The aircraft bore U.S. civil registrations and airline livery. (Ministerio de Defensa Nacional).

delivered to Bolivia and that they would probably end up costing around $75,000 each, instead of $30,000."

It is also worth noting that Paraguay had, at some point, entertained the idea of replacing her tried and true Potez 25s with another multi-purpose aircraft from the same firm, the Potez 50. A total of nine aircraft of this type has been suggested but, again, Paraguay's financial situation precluded such an order.

Bolivia's preoccupation with obtaining bombers led to one other abortive sales initiative circa May 1933 suggested by the Curtiss-Wright Export Corporation representatives in La Paz. In a communication with the home office, "Cliff" Travis, while passing on the generally laudatory comments of the Bolivians regarding their "Hawks" and "Ospreys" also advised that some of the Bolivian officers felt that these were "mere toys, because they could carry only some 460 pounds of bombs....which seems to be the natural reaction of these people as, after all of these bombing expeditions with really good results, they have gone bomb-minded." In this same letter, he suggested that, in view of the Bolivians' avowed interest in bombers with larger capacities, that "there may also be a chance to sell some Travel Air bombers," although it is unclear exactly what types he had in mind at the time.

While the reader will have obviously noted the dominance of the Curtiss-Wright Export Corp. in managing to sell significant numbers of aircraft to, especially, Bolivia, it should not be forgotten that the chief U.S. competitor to Curtiss-Wright, the United Aircraft Corporation, was also engaged in aircraft and arms sales into South America at this point, having scored orders in Peru, Brazil and Argentina in particular for aircraft products of two of their principle subsidiaries, Douglas and Vought. By September 1933, United was actively engaged in attempting to sell Curtiss and Curtiss-Wright aircraft to Bolivia in competition with Curtiss-Wright Export, the difference being that United was attempting an "end run" on the Curtiss efforts in La Paz by engaging the support of none other than General Kundt, Colonel Bilbao and the (then) Minister of War, Hertzog. Indeed, had it not been for the active intervention of Bolivian pilots themselves, history may have recorded that Bolivia went to war in the Chaco with a combination of Douglas, Vought and United-brokered Curtiss-Wright aircraft!

6

The Aftermath

When the armistice was finally placed into effect on 12 June 1935, the air arms of both nations were practically exhausted, not to mention the down-trodden and long-suffering ground forces.

Both air arms had suffered considerably during the war and losses, both to personnel and equipment, in both actual operations and routine training, had been tremendous.

However, as is so often the case, although they could ill afford it, both nations almost immediately set about seeking parts needed to repair and maintain their existing equipment and, at the same time, more efficacious and modern types with which to replace them, should hostilities once again break out. It should be recalled that, although an Armistice had been agreed upon in 1935, an actual settlement of the war and a peace treaty were not signed until 1939.

After the war, both Bolivia and Paraguay set about rearming in anticipation of a renewal of hostilities. Although both were financially strapped - especially Paraguay - some modern aircraft were acquired, regardless of the expense. One type sought by Paraguay was the odd looking Caproni Ca 309 "Ghibli" bomber/transport. This is T-5, named "Mariscal Jose F. Estigarribia" at Ñu-Guazu in 1942. (Archivo Cnel. Av. (S.R). A. Pasmor).

Acquired in late 1938, Paraguay originally ordered three Ca 309s, but only two were actually delivered. By 1945, these had passed to the use of the Army's airline, L.A.T.N. and the military serials had been dropped, but not the names. This is c/n 372 "Mariscal Jose F. Estigarribia." (Antonio Sapienza)

Paraguay, in particular, seems to have been acutely aware of the lessons learned during the war, and how far worse things may have gone for her had her gallant ground forces not proven more effective in the "Green Hell." Bolivia, on the other hand, felt very strongly that she had "won" the war in the air, and that her forces need only be strengthened along certain very particular lines.

German, Italian and, to a lesser extent, U.S. firms soon became aware of the interest of both nations in rearming in the air. One of the most pressing needs was for dedicated, and relatively modern trainers, and apparently both nations displayed an interest in the fairly efficient Italian Breda Ba 25 primary trainer biplane. Bolivia evinced an interest in at least six of these aircraft as part of a package which would also include a number of Breda Ba 65 attack aircraft but, when it was learned that Italian agents were also attempting to sell the exact same package to Paraguay, the negotiations were terminated peremptorily. Paraguay, for her part, did indeed acquire four Ba 25s via the Italian broker CIMA (Consorzio Italiano de Materiale Aeronatuiche), three of them for the Fuerzas Aerea Nacionales and one, on floats, for the Navy. Two of these were standard primary trainers costing $30,760.00, one a dedicated blind flying trainer (at $17,197.00) and the fourth a float-equipped example for the Navy (at $19,197.00.) Ordered, apparently sometime in 1937, these did not arrive until some time later. The Army aircraft reportedly gained serial codes E.1, E.3 and E.5, while the Navy aircraft apparently also carried serial "1" (sometimes given also as E.1). One of the Army aircraft survived into January 1949, while the Navy "Ba 25 Idro" with a 240hp Alfa Romeo D2 engine, was still on the Order of Battle as of January 1947.

As mentioned earlier, Italy also attempted to interest both Bolivia and Paraguay in the monoplane Breda Ba 65 attack aircraft, although Bolivia ultimately bowed out of the negotiations at the 'interest' stage. Paraguay apparently negotiated at some length for variants of the Ba 65bis, but couldn't arrive at mutually agreeable terms, and the idea was finally dropped altogether, contrary to some reports.

Paraguay also recognized the need for a strategic bombing force, the specter of which had hung over their capital during practically the entirety of the war. At one point, Paraguay had approached the Italian consul in Asunción with a proposal to acquire "20 to 30 Caproni Ca 101s with which to bomb La Paz." However, the international family of nations was by then publicly opposed to arming the belligerents, and the Italians declined to become further involved. Consequently, when the negotiations for the Ba 65s also broke down,

An extraordinary, and unlikely 1940 lineup of virtually the entire Paraguayan air arm features a rare Brazilian donated Muniz M-9 trainer, both Caproni Ca 309s, three Fiat CR.32s, one Fiat CR.30 and seven Caproni/Bergamaschi AP.1s! (Archivo FAP via Cnel. Ugarte).

On paper, at least, the post-Chaco War Paraguayan air arm was formidable, and included the two-place Caproni/Bergamaschi AP.1 attack aircraft. Delivered in 1937 and serving into 1947 with the Arma Aerea, this 1942 view shows the unusual camouflage scheme carried during a goodwill flight to Argentina. (Antonio Sapienza).

between May and 19 October 1937, Paraguay requested a citation - again as part of a larger package - for 11 "Caproni single engined bomber/transports." These, almost certainly Ca 111s, also came to naught as, again, terms could not be agreed upon. At the same time, however, similar interest was expressed in 12 "Caproni two-place, all purpose twins with 740hp engines," but again, when the associated costs and terms were announced, coupled with a change in Government in Paraguay, this proposal also was aborted. However, apparently Paraguay had put some "up front" money into the overall Caproni negotiations under the previous government and, as a result, a settlement of the account was finally achieved by delivering two (of three) Caproni Ca 309 "Ghibli" light bomber/transports in mid-1939 valued at $126,500. Coded in the transport category as T-3 "Mariscal Lopez" and T-5 "Mariscal Estigarribia," these were accompanied by seven Bergamaschi (Caproni) AP-1 monoplane (although up to 26 of these had originally been discussed), single-engined fighter/attack aircraft, valued at $273,300. These arrived in 1939 also, and were coded 2.1, 2.3, 2.5, 2.7, 2.9, 2.11 and 2.13, some surviving in service into January 1949. Thus, at last, Paraguay had a mix of relatively modern, competent combatant types with which to face Bolivian counter moves. The AP-1s were assigned to Paraguay's Segunda Escuadrilla de Caza.

Having taken care of her limited bomber/transport and attack needs, Paraguay had also initiated measures to resurrect her all but non-existent fighter force. Turning once again to Fiat (in spite of its mixed experiences with the CR.20s), she ordered five classic Fiat CR.32quater fighter biplanes. Indeed, after the Italian CIMA agent in Asunción, Francini (and another agent in Buenos Aires, one Sr. Pastega), had negotiated the semi-aborted large-scale Caproni contract with the aid of the Italian Minister at the time, Mariano de Angelis, totaling some $598,984.00., the Paiva Government of Paraguay desired to rescind this prohibitively expensive agreement. However, after much insistence on the part of the Italians, it was finally agreed to continue on to the amount of $337,829.63, of which the cost of the five CR.32s (($187,500.00) was part. These apparently arrived in Asunción in 1939 also, although some sources claim 1938 (which seems unlikely) and were coded 1.1, 1.3, 1.5, 1.7 and 1.9. These were assigned to the Primera Escuadrilla de Caza.

The AP.1 was a fairly large aircraft, and the first truly modern aircraft operated by the Arma Aerea Paraguaya. The only concession to the upper surface camouflage scheme was the national colors on the rudder. (Hagedorn Collection).

It is not generally known that, along with the CR.32s, Paraguay also acquired a pair of Fiat CR.30B two-seat fighter/trainers at a cost of $57,000. Frequently cited as having been delivered during the conflict (some sources state as early as November 1934), this did not happen, and the two CR.30Bs were given serials E.7 and E.9, one surviving in service as late as August 1944.

Bolivia, on her own part, had not been idle. Although the negotiations with the Italian agency had been abortive, Bolivia sought to renew her contacts with German and U.S. sources, these suppliers having provided aircraft which had, for all intents and purposes, won the air war for her previously.

Accordingly, one of the earliest post-Armistice acquisitions was, like Paraguay, for much needed trainers, and three Klemm Kl 25 monoplane primary trainers were acquired by mid-October 1936. Supplied with 80hp Hirsch engines, these were assigned to the Escuela de Aviación Militar "Mayor Pabon" at Cochabamba, where they served alongside the miscellany of left-over Chaco War veterans at least into 1938.

Before the ink on the armistice agreement was even dry, however, circa October 1935, German sources had offered Bolivia terms on a number of Junkers Ju 160 single-engined transport/bomber monoplanes, but apparently Bolivia did not have the foreign exchange or energy to invest in such a purchase at that point, and, perhaps equally, was anxious not to "upset the apple cart" in the peace process.

Bolivia did invest modestly in German built equipment starting as early as 17 November 1937, however, when a single Focke-Wulf Fw 44M trainer biplane was demonstrated to Bolivian authorities and subsequently acquired to add to the strength of the hard-pressed EAM. The former German registered D-EUGO, this aircraft was unusual in having a 150hp Menasco engine in Bolivian service and, as of 1 October 1940, was the only airworthy trainer at Cochabamba, and the only FAB aircraft not at La Paz. It was scrapped sometime between then and December 1940 after suffering an accident.

Along with the earlier offer of Ju 160s to Bolivia in October 1935, the German agents had also suggested the twin-engine Ju 86 bomber/transport. While this offer was not acted upon at the time, Bolivia subsequently did order a total of four Ju 86s, consisting of two Ju 86Es for the FAB as bomber/transports and two Ju 86Z transports for LAB. The bomber/transports were numbered "1" and "2" and named "Monasterios" and "Beltran" and arrived in 1937, while the two transports were named "Illimani" and, later, "Mariscal

While the Caproni/Bergamaschi AP.1s looked and sounded good, considerable trouble was experienced with their engines. Here (then) Captain A. Pasmor poses before one of the maintenance nightmares in 1939 at Ñu-Guazu. (Archivo Cnel. Av. (S.R). A. Pasmor).

Dedicated trainers were one of the chief needs of the post-war Paraguayan air arm. As part of the overall Italian arms package, at least three Breda Ba 25 trainers were acquired for the air force and others for the Navy. This photo shows the President of Paraguay, Dr. Felix Paiva (third from left) and other high officials beside a new Ba 25 in 1939 at Ñu-Guazu. (Archivo Cnel. Av. (S.R). A. Pasmor).

Sucre," "Mariscal Santa Cruz" and "General Perez." In truth, all but one of the Bolivian Ju 86s apparently had nose gun positions and were convertible to bombers, and these aircraft became pawns in the on-going peace negotiations with Paraguay. Eventually, the LAB aircraft were passed to the FAB between 1 April 1938 and 1 January 1939, although the third aircraft remained with the airline until 23 January 1942 (probably "Illimani"). Two were lost in accidents on the same day (7 March 1943) and only two remained by 4 September 1944, although both were grounded for lack of needed parts.

Yet another odd German aircraft was acquired for Bolivia in 1938, a single example of the elegant Focke-Wulf Fw 56A "Strosser," the former German civil D-IBAE. Demonstrated to the Bolivians at the same time as the Fw 44M, the Fw 56A survived until September 1943, by which time it was reported in "vcry bad condition at Cochabamba."

The Paraguayan fighter arm was also given new blood in the form of five Fiat CR.32quaters. These were camouflaged similarly to the Caproni/Bergamaschi AP.1s. Here, six Paraguayan fighter jocks pose before one of the new mounts at Ñu-Guazu in 1940. (Archivo Cnel. Av. PAM F. Ugarte).

Bolivia was not idle following the war. While she emerged with a considerable number of effective aircraft types still intact, most of these were nearly worn out, and the Bolivians neglected to seek replacements for them until it was almost a crisis. When they did, they countered known Paraguayan acquisitions by purchasing 10 of the classic Curtiss-Wright 19R all-metal monoplanes from their old trading partner in June 1938. Here, one of these flies over the Andes with a war veteran "Cyclone Falcon," neither of which bear serial numbers. (Ramiro Molina Alanes).

But unquestionably one of the most important single purchases made by Bolivia during the post-Armistice period was for 10 Curtiss-Wright 19R light combat aircraft which, after rather protracted negotiations, finally arrived on delivery on 29 June 1938. These were given Bolivian serials 118 to 127 but, following arrival, were not finally erected and test-flown until 12 October 1939. These aircraft were viewed as the best aircraft for the money to counter the Bergamaschi AP-1s and Fiat fighters acquired by their former antagonist and, indeed, Bolivia got her moneys worth from the classic 19Rs. Some remained in service as late as 1955, and one still exists today, and by the time this is being read will have been returned to the United States for full restoration.

The final aircraft acquired by Bolivia during the post-war Armistice leading up to the final peace treaty was a single Ryan STA Special c/n 199, NC18905, which arrived in Bolivia 15 January 1939, although it was not sold to the FAB until a short time later. Interestingly, it is thought that this single Ryan was the aircraft which gave rise during the years that followed to the reports that Bolivia had acquired examples of the "Ryan PT-16A" under "Lend-Lease."

Paraguay also acquired two two-seat Fiat CR.30B fighter/trainers in 1939, and Captain PAM Fabio Martinez poses here in front of one of these at Ñu-Guazu in 1940. (FAP).

Another Bolivian purchase, and one which struck fear into the hearts of Paraguayans, was a mix of three Junkers Ju 86B-0 and Ju 86K-7 bomber/transports acquired in 1938. Although bearing joint military and civil (LAB) markings, they were every inch bomber capable aircraft. Here is s/n 2 at Santa Cruz in 1940. (Hagedorn Collection).

The hero of the Chaco War, President (General) Jose Felix Estigarribia is in the white suit in the midst of this group of Army and Navy officers at Ñu-Guazu in 1940. The two Fiat CR.32quaters in the background reveal some markings details. (Antonio Sapienza).

Sources

PRIMARY SOURCES

Library of Congress, Washington, DC
One of the most important sources utilized, especially with regard to Curtiss-Wright Export Corporation activities during the Chaco War, exists in the Law Library Reading Room, where transcripts of Congressional Investigations of the mid to late 1930s into the Arms Industry are located on microfiche. An index for these, by broad general subject, is available there.

National Archives, College Park, MD
Primarily Records Group (RG) 59, Records of the Department of State, for the periods 1919-1940. Series consulted included Aircraft, Exports, Arms Transfers, Military Affairs for not only Bolivia and Paraguay, but also for Argentina, Chile and Peru. Many of these were found on Microfilm in the Microfilm Reading Room on reels M0644 and others. RG 18, Records of Headquarters, U.S. Army Air Forces, Central Decimal Files 1917-1938 (and predecessor organizations) were also extensively consulted, especially War Department General Staff, Military Intelligence Division (G-2) attache reports (Foreign Aviation Reports 360.02 et al) and miscellaneous reports from the same countries.

PERSONAL INTERVIEWS

Coronel Agustin Pasmor, former Chief Maintenance Officer of the Paraguayan Air Arm during the Chaco War.
Lieutenant Gonzalo Palau, a Potez 25 gunner/navigator during the Chaco War.
Captain Abdon Alvarez Albert, a Fiat C.R.20/30/32/Caproni AP-1 pilot during and after the Chaco War.

BOOKS AND PERIODICALS

Alvarez Albert, *CAP (RVA) P.A.M. Abdon Con las llamas en el Aire*, Imprenta Militar, Asunción, 1979.

Andrews, *C. F. Vickers Aircraft Since 1908*, Putnam, London, 1969.

Aponte, Mayor (S.R.) *P.A.M. Leandro Cincuenta Años de Aeronáutica en el Paraguay*, El Arte, S.A., Asunción, 1957.

Aponte, Mayor (S.R.) *P.A.M. Leandro La Aviación Paraguaya en la Guerra del Chaco*, Imprenta Militar, Asunción, 1985.

Arze Quiroga, E. ed. *Documentos para una historia de la guerra del Chaco*, seleccionados del archivo de Daniel Salamanca, La Paz, 1951.

Ayala Moreira, *Rogelio Por que no ganamos la Guerra del Chaco*, La Paz, 1959.

Baptista Gumucio, *Mariano Historia (Gráfica) de la Guerra del Chaco*, Editorial Khana Cruz S.R.L., La Paz, 1982.

Bowers, Peter M. *Curtiss Aircraft 1907-1947*, Putnam, London, 1979.

Bozzano, Cap. de Nav. (S.R.) *Ing. Naval Jose Reminiscencias*, Casa Editora Toledo, Asunción, 1962.

Centurion, Carlos *El conflicto del Chaco Boreal*, gestiones diplomaticos, La Colmena, Asunción, 1937.

Cespedes, A. *Sangre de mestizos: relatos de la Guerra del Chaco*, La Paz, 1983.

Costagliola, Antonio *Histórico vuelo Buenos Aires-Asunción*, Ed. Trinchera, Asunción, 1973.

De Los Santos, Tomas *La Revolución de 1922*, Tomos 1 y 2, Editorial El Lector, Asunción, 1984.

Diaz Machicao, *Porfirio Historia de Bolivia: Salamanca, la guerra del Chaco*, Tejada Zorzano, 1931-1936, Gisbert, La Paz, 1955.

English, Adrian J. *Armed Forces of Latin America*, Jane's, London, 1984.

Estigarribia, Jose Felix *The Epic of the Chaco: Marshal Estigarribia's memoirs of the chaco War., 1932-1935*, University of Texas Press, Austin, 1950.

Fernandez, Jose Carlos *La guerra del Chaco*, Buenos Aires, 1955-v.

Garner, William R. *The Chaco dispute, a study of prestige diplomacy* by William R. Garner, Public Affairs Press, Washington, 1966.

Gonzalez, Tcnel (S.R.) *Antonio E. Preparación del Paraguay para la Guerra del Chaco*, Tomos 1 y 2, Editorial El Gráfico S.R.L., Asunción, 1957.

Guachalla, L. F. *Jayucubas Comentarios y cronicas de la Guerra del Chaco*, Amigos del Libro, La Paz, 1978.

Guanes Machain, Luis *Recuerdos y observaciones de la Guerra del Chaco*, El Arte, Asunción, 1963.

La Foy, M. *The Chaco dispute and the League of Nations*, Bryn Mawr, 1941.

Lara Castro, Mariano Luis *La Preparación de la Defensa Nacional y la Conducción Militar y Diplomática de la Guerra del Chaco*, Anuario de la Academia Paraguaya de la Historia, Volumen XIX, Asunción, 1982.

Lieuwen, Edwin (Ed.) *Arms and Politics in Latin America*, Council on Foreign Relations, Praeger, New York, 1961.

Mercado Moreira, M. *Historia diplomatica de la Guerra del Chaco*, La Paz, 1966.

Ortiz Pacheco, Nicolas *La Justicia contra el machere*, La Paz, 1935.

Pasmor, Cnel Av. (S.R.) Agustin *Creación de la Aeronáutica Militar Paraguaya*, Trinchera, Asunción, 1975.

Pasmor, Cnel (Av) (S.R.) Agustin *Origen, Organización y Desarrollo de la Actividad Aeronáutica en el Paraguay*, Trinchera, Asunción, 1985.

Paz Soldan Pol, Gral.Div. Ae.(SP) *Conduccion de la Fuerza Aerea Boliviana en la Guerra del Chaco*, Editorial Aeronáutica de la FAB, 1990.

Perez Acosta, E. *La contienda del Chaco*, lo que vio, oyo y supo un capellan, 1962.

Pol, H. R. *La Campana del Chaco: glosas y reflexiones militares*, 1945.

Querejazu Calvo, Roberto *Masamaclay, historia politica, diplomatica y militar de la guerra del Chaco*, La Paz, 1965.

Ramirez, J. I. *La paz del Chaco, la defensa de la linea de Hitos y el comite de los tres*, Buenos Aires, 1942.

Rios, Angel *La defensa del Chaco, verdades y mentiras de una victoria*. Buenos Aires, 1950.

Rojas Gonzalez, Cap. Nav. (S.R.) Agustin *Algo sobre la contribución logística de la Armada Nacional en la Guerra del Chaco*, Imprenta Militar, Asunción, 1985.

Rout, Leslie B. *Politics of the Chaco Peace Conference 1935-39*, Austin, 1970.

Scheina, Robert L. *Latin America: A Naval History, 1810-1987*, Naval Institute Press, Annapolis, 1987.

Seiferheld, Alfredo *Conversaciones politico-militares*, Tomo IV, Ed. Histórica, Asunción, 1987.

Seiferheld, Alfredo *Economía y Petroleo durante la Guerra del Chaco*, Ed. El Lector, Asunción, 1983.

Siegrist, Martin E. *Bolivian Air Power - Seventy Years On*, AIR International, Volume 33, Number 4, October 1987.

Sosa Valdez, Mayor M.A.M. (S.R.) Raul Agustin *Acciones de la Aviación en Campaña en la Guerra del Chaco*, Trinchera, Asunción, 1975.

Stagni, Cnel D.E.M. (S.R.) Pablo *El Ejército y la Historia*, Tomo 1, Imprenta Militar, Asunción, 1986.

Stefanich, J. *El 23 de octubre de 1931, primera batalla por la defensa del Chaco y primer grito de la revolucion de febrero de 1936*, Buenos Aires, 1959.

Torres Ortiz, H. *Campo Via, antecedentes y consecuentes*, Fenis, La Paz, 1937.

Valdovinos, A. *Bajo las botas de una bestia rubia*, 2.ed, Asunción, 1933.

Vergara Vicuna, A. *Historia de la guerra del Chaco*, La Paz, 1940-44.

Villa de la Tapia, *Tcnel Av. Amalia Alas de Bolivia*, Tomo 1, 2, 3, Ed. Aeronáutica, La Paz, 1983-

Wewege-Smith, Thomas *War planes and women, the enthralling story of an airman's adventures in love and war*, Hutchinson, London, 1938.

Zarate Monges, Cap. P.A.M. (S.R.) Felix *La Aviación Paraguaya antes y durante la Guerra del Chaco*, Offset Leguizamón, Asunción, 1988.

Zook, D. H. *The conduct of the chaco War*, New York, 1961.

Zook, CPT David H., Jr. and CPT Crabbe, William M., Jr., *Airpower in the chaco War*, Air Power Historian, 1963.